Nature-Inspired Self-Cleaning Surfaces in the Nanotechnology Era

Edited by Phuong V. Pham

Published in London, United Kingdom

Nature-Inspired Self-Cleaning Surfaces in the Nanotechnology Era
http://dx.doi.org/10.5772/intechopen.106087
Edited by Phuong V. Pham

Contributors
Shahnaz Mansouri, Mojdeh Nakhaei, Jing Ying Chong, Yunlong Tang, Deepanjana Adak, Raghunath Bhattacharyya, Thi Viet Ha Tran, Minh Viet Nguyen, Le Minh Tri Nguyen, K.S. Srin, J. Ramkumar, Ravi N. Bathe, Shreedhar Bhat, Roopali Rai, Kamlesh Panwar, Neelam Vyas, Abdullah Al Nahid

First published in London, United Kingdom, 2023 by IntechOpen
IntechOpen is the global imprint of INTECHOPEN LIMITED, registered in England and Wales, registration number: 11086078, 5 Princes Gate Court, London, SW7 2QJ, United Kingdom

British Library Cataloguing-in-Publication Data
A catalogue record for this book is available from the British Library

Additional hard and PDF copies can be obtained from orders@intechopen.com

Nature-Inspired Self-Cleaning Surfaces in the Nanotechnology Era
Edited by Phuong V. Pham
p. cm.
Print ISBN 978-1-83769-069-5
Online ISBN 978-1-83769-070-1
eBook (PDF) ISBN 978-1-83769-071-8

Meet the editor

Phuong V. Pham is an assistant professor in the Department of Physics, National Sun Yat-sen University (NSYSU), Taiwan. He is a pioneering scientist in materials science and electronics/optoelectronics. He has participated in industrial projects with Samsung, POSCO, the Government of Korea, the Ministry of Education and the Ministry of Science and Technology of China, and the National Science and Technology Council (NSTC), Taiwan. He was formerly a senior scientist at the Hangzhou Global Scientific and Technological Innovation Center (HIC), Zhejiang University, China. He earned a Ph.D. from SKKU Advanced Institute of Nanotechnology (SAINT), Sungkyunkwan University (SKKU), South Korea. Then, he spent a few years as a postdoctoral researcher and research fellow at the School of Advanced Material Science and Engineering, SKKU, and at the Center for Multidimensional Carbon Material (CMCM), Institute for Basic Science (IBS), South Korea, respectively. He is a recipient of the National Science Foundation (NSF) Career Award, the National Postdoctoral Award for Outstanding Young Scientists, China (2019), and the NSTC Grant Award (2023). He is a book editor and has published more than fifty peer-reviewed papers, books, and book chapters. His research interests include materials physics, surface physics, atomic film syntheses, 2D heterostructures, donors/acceptors, nanocomposites, block copolymers, and plasma engineering for electronics/optoelectronics.

Contents

Preface

The 21st century is the age of nanotechnology, a rapidly developing field related to almost every area of life.

Nature-inspired, self-cleaning surfaces have attracted attention for both fundamental research and practical applications. Self-cleaning ability on technical surfaces can enhance the outstanding values of products in a variety of promising applications, such as microfluidics, lap-on-chip, microreactors, air purification, anti-microbial activity, and so on. A conventional path to obtaining self-cleaning properties is creating superhydrophobic surfaces so water droplets can roll down, picking up dirt particles. This book discusses nanomaterials-based nanotechnology and chemical engineering, focusing on mechanisms, modeling, and manufacturing nature-inspired, self-cleaning surfaces. It addresses the latest advancements in self-cleaning surfaces of materials and discusses how these materials can be created, focusing on scalable production. This book will strengthen our understanding of nature-inspired, self-cleaning surfaces and stimulate interdisciplinary collaborations across nanotechnology, materials science, chemistry, biology, and engineering.

Chapter 1 introduces various biomimetic superhydrophobic materials, their fabrication approaches, and their promising applications in environmental treatment. The chapter presents the overall research directions on biomimetic superhydrophobic materials.

Chapter 2 highlights the various self-cleaning surfaces of plants such as lotus, rice, Indian canna, taro, cabbage, Indian cress, *Salvinia molesta*, pitcher, *Anubias barteri*, and *Heliamphora nutans*. It provides an in-depth understanding of the mechanisms and assembling of these self-cleaning plant surfaces and their industrial applications in the food, pharmaceutical, and other industries.

Chapter 3 discusses the various mechanisms for self-cleaning, including super-hydrophobicity, super-hydrophilicity, and photocatalysis,, emphasizing polyurethane origin. Additionally, the chapter presents applications of such polyurethane surfaces, such as anti-fogging, anti-icing, anti-reflection, corrosion resistance, drag reduction, sensors, solar cells, and textiles.

Chapter 4 highlights the use of metal oxide-based nanocomposites for self-cleaning purposes. It provides an outlook of different metal oxide and metal–metal oxide nanocomposites in advancing self-cleaning properties, durability, and other mechanical properties. It also presents a variety of polymeric metal oxide-based systems and methods and their related mechanisms for enhancing self-cleaning behavior.

Chapter 5 provides an overview of plasma and cold-plasma generation and the self-cleaning approaches assisted by plasma gases and related technologies such as plasma cleaning, plasma etching, plasma polymerization/deposition, and so on. Moreover, it also discusses the exciting combination of plasma and sol–gel approaches.

Chapter 6 analyzes laser technology-induced periodic surface structures on superhydrophobic self-cleaning applications. The method with physical alteration will give the durability of hydrophobic properties on the material surface, and the alteration of this property is based on the machining process. The method's fast process for surface alteration is its primary advantage in self-cleaning applications. Suitable materials with specified textures may provide a foundation for self-cleaning surfaces for food processing components and equipment. It is a speedy method for the development of multifunctional surfaces. Hydrophobicity with super or ultra-featured contact angle shallow sliding or rolling off-angle can instrument self-cleaning physical activities in solar panels, hydraulic equipment, and agri machines for instant residue removal. In addition, to prevent ice formation on power transmission cables and aircraft wings and cockpit shields and corrosion resistance of structures, water pipelines, hydraulic components, and other machinery in addition to biofouling resistance on ship basements structures and other pipelines and thermal barrier coating, enhanced broadband absorption.

As an editor, I wish to acknowledge all the authors for their excellent contributions. I also wish to thank Publishing Process Manager Maja Bozicevic and Commissioning Editor Sandra Bolf at IntechOpen for their assistance throughout the publication process.

Phuong V. Pham
Department of Physics,
National Sun Yat-Sen University,
Kaohsiung, Taiwan

Chapter 1

Biomimetic Superhydrophobic Materials for Environmental Applications

Thi Viet Ha Tran, Minh Viet Nguyen and Le Minh Tri Nguyen

Abstract

Environmental pollution has been one of the people's most significant concerns for decades. In today's industrialized and modernized society, the problem of environmental pollution has become more and more serious, directly affecting the sustainable development of each country. The unique surface properties of materials and interfaces produced by biomimetic approaches can be leveraged to create practical solutions to challenging environmental issues. Among them, superhydrophobic materials get a lot of attention because of their exceptional capacities in various environmental applications such as oil-water separation, membrane-based water purification and desalination, biofouling prevention, high-performance vapor condensation, and atmospheric water capture. This chapter reviews and discusses the fundamental principles of superhydrophobicity, recent works in preparing superhydrophobic surfaces, their potential environmental applications, and the challenges confronted in their new applications.

Keywords: superhydrophobic, biomimetic, environment, pollution treatment, application

1. Introduction

Nature inspires many scientists and engineers to create remarkable inventions for human life. Humans observe and imitate numerous natural materials, structures, and systems to design and invent new products. This creative process is known as biomimetic. One notable example of biomimetics is the superhydrophobic surface. In nature, we can find many superhydrophobic surfaces of plants and animals, such as lotus leaves and butterfly wings, with waterproof, self-cleaning, and anti-adhesion properties. These properties are studied and applied in many fields of life, such as making anti-fogging materials [1], anti-freezing materials [2], self-cleaning materials [3], or environmental treatment [4].

Recently, superhydrophobic surfaces with high applicability have received more attention from researchers. Various methods have been developed to prepare superhydrophobic surfaces, for example, the sol-gel method [5], self-assembly [6], chemical etching [7], plasma etching [8], vapor deposition [9], and so on. Moreover, there are also many studies on fabricating superhydrophobic surfaces on different substrates such as fabric [10], glass [11], silicon [6], or metal surface [5, 12] with diverse applications. The appropriate fabricating method will be chosen and developed differently based on the substrate to prepare the superhydrophobic surface successfully.

Since the nineteenth century Industrial Revolution to this day, pollution has increased quickly, directly impacting the ecosystem, including animals, plants, and human health and life. Several materials have been researched in order to deal with environmental issues. Among them, superhydrophobic materials with a unique behavior against water droplets received much attention from researchers, and they have become widespread and popular in recent years. This chapter reviews the basic contact angle science in the first part. The fabrication methods and applications of superhydrophobic materials in environmental treatment are overviewed in the following section. Some crucial issues affecting the unsuccessful wide-range applications of superhydrophobic surfaces are addressed critically in the conclusions and outlook section. Finally, some proposals are put forward for future guidance on the environmental applications of superhydrophobic surfaces.

2. Superhydrophobic surfaces in natural

Nature inspires many scientists and engineers to create remarkable inventions for human life. Humans observe and imitate numerous natural materials, structures, and systems to design and invent new products. This creative process is known as biomimicking. Biomimicking refers to designs that emulate or imitate nature's models, systems, and elements to solve complex human problems. It aims to draw inspiration from nature's engineering to solve the world's most pressing challenges and ensure a sustainable future for all life. Superhydrophobic surfaces are an essential aspect of biomimicking.

Water contact angle (WCA) is the parameter used for quantifying the wettability of a surface. The contact angle is defined as the angle created by the intersection of the liquid–solid interface and the liquid-vapor interface (geometrically obtained by making a tangent line from the contact point along the liquid-vapor interface in the droplet profile) [13]. In 1805, Young proposed the first fundamental equation that quantified the hydrophobicity/hydrophilicity of a surface based on a static contact angle [14]:

$$\cos\Theta = \frac{\gamma_{SV} - \gamma_{SL}}{\gamma_{LV}} \tag{1}$$

in which θ is the static water contact angle, γ_{SV}, γ_{SL}, and γ_{LV} are the interfacial tensions of the solid-vapor, solid-liquid and the liquid-vapor interface, respectively. In presenting this relational equation, he is widely considered to be the pioneer of scientific research on the wettability and water contact angle.

A surface is considered hydrophilic when WCA <90 degrees ($\gamma_{SV} > \gamma_{SL}$), while it is considered hydrophobic when WCA ≥ 90 degrees ($\gamma_{SV} \leq \gamma_{SL}$) [13]. The pictures of hydrophilic and hydrophobic surfaces are shown in **Figure 1**.

However, Young's equation is only applicable to solids that are perfectly smooth and chemically homogeneous. The wettability of rough or chemically heterogeneous surfaces, which are more suitable, is much more complicated. Two remarkable models

DOI: http://dx.doi.org/10.5772/intechopen.111444

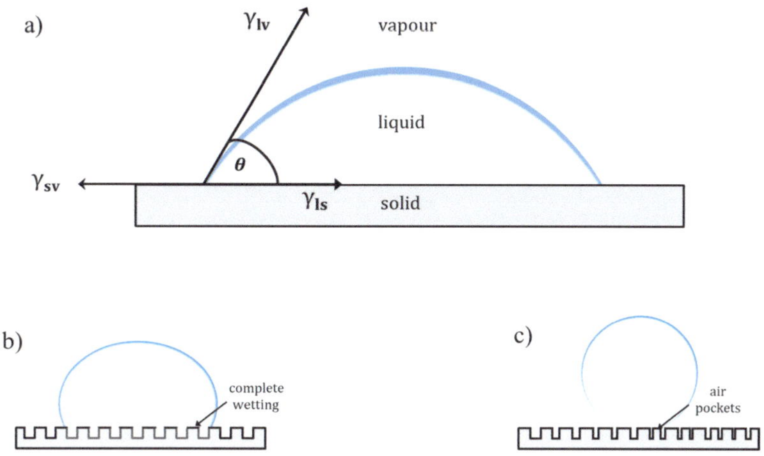

Figure 1.
Different states of the wetting behavior of solid surface.

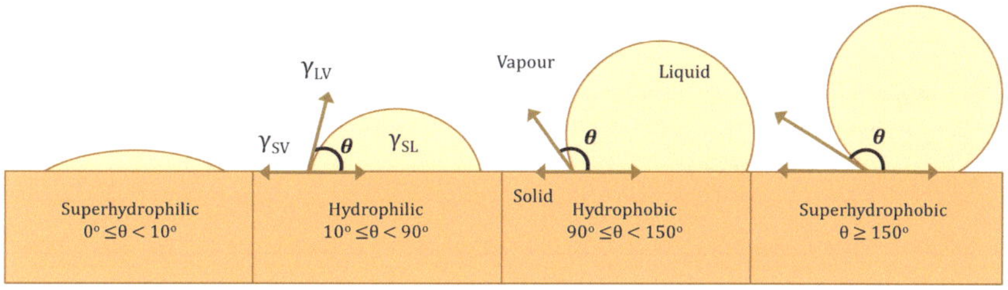

Figure 2.
Schematic of (a) Young's equation, (b) Wenzel's model, and (c) Cassie-Baxter's model.

describing the impact of surface roughness on water contact angle are the Wenzel model and the Cassie & Baxter model (**Figure 2**).

In 1936, Wenzel developed a model in which the water drop can penetrate the grooves on a rough surface. Over the course of his experiments, he discovers that roughness makes a hydrophilic surface more hydrophilic and makes a hydrophobic surface more hydrophobic. He proposed an equation showing the relationship of the water contact angle on the smooth surface and rough surface as follows:

$$\cos\Theta = \frac{r(\gamma_{SV} - \gamma_{SL})}{\gamma_{LV}} = r\,cos\Theta \tag{2}$$

In which $\theta^{\wedge *}$ is the water contact angle on a rough surface, θ is the water contact angle on a similarly smooth surface, and r is the surface roughness factor, defined as the ratio of the actual area of the solid surface to the projected area (r = 1 for a perfectly smooth surface, and r > 1 for a rough one). The Wenzel equation indicates that the hydrophilicity is enhanced by roughness when θ is <90 degrees, whereas the hydrophobicity is increased by roughness when θ is >90 degrees.

After that, in 1944, Cassie and Baxter reported that Wenzel's equation could not accurately predict the water contact angle of droplets on rough surfaces with air pockets trapped in the rough grooves. Consequently, they proposed an alternative model in which the wetting state is considered: The grooves under the droplet are filled with vapor instead of liquid. They modified Wenzel's equation and presented another one that can predict the contact angle in these cases as follows:

$$\cos\Theta = f_1 \cos\Theta_1 + f_2 \cos\Theta_2 \tag{3}$$

in which,

$$f_1 = \frac{\text{area in contact with droplet}}{\text{projected surface area}} \tag{4}$$

$$f_2 = \frac{\text{area in contact with air beneath the droplet}}{\text{projected surface area}} \tag{5}$$

and θ_1 represents the contact angle of the smooth solid surface, and θ_2 represents the contact angle for air.

A superhydrophobic surface can be defined as the surface which have the high water contact angle (≥150degrees), low sliding angle/shedding angle (≤10 degrees), anti-sticking, anti-contamination, and self-cleaning effect. The superhydrophobic surface can be found in both natural plants and animals.

The best-known example of superhydrophobic natural surfaces is lotus leaves (*Nelumbo nucifera*), characterized by θ >150 degrees, very low water adhesion and self-cleaning properties. The self-cleaning capacity of directly removing dust and particles by displacing water droplets is derived from the Cassie–Baxter state. This property arises the from the dual (micro/nano) surface structure of Lotus leaves. Besides, in nature, numerous of plants like rice leaves and rose petals carry out the excellent hydrophobicity or superhydrophobicity. In the case of animals, insect wings' superhydrophobicity is an advantage in reducing dust/particle contamination and improving their flying capacity. Different families with highly hydrophobic wings, including dragonflies mayflies, stoneflies, lacewings, alderflies, caddisflies, butterflies, moths, etc. ... have been founded. Many insects' feet are also superhydrophobic. For instance, geckos are able to climb on vertical surfaces thanks to their feet with well-aligned microscopic hairs known as setae.

Nature's mimicry is the easiest way to replicate materials with similar properties because nature has produced plants, insects, and animals that capable of repelling water over millennia. The properties of superhydrophobic surfaces are the potential for many real-world applications like the production of anti-fogging materials, anti-freezing materials, and self-cleaning materials.

3. Fabrication of superhydrophobic materials

3.1 Chemical etching method

Chemical etching is a wet method in which the surface molecules react with highly acidic or basic solutions to produce roughness. Although it only requires simple equipment and simple chemical at a low cost, it is a fast process and provides a high etching rate and selectivity. However, it still has some disadvantages. This method requires considerable amount of etching chemicals and can cause contamination to the substrates. It is also difficult to control the etching rate of the process.

Many researches using this method to fabricate superhydrophobic surfaces have been published. For example, in Qian and Shen's studies [7], the surface of aluminum, copper, and zinc surface was chemically etched with Beck's dislocation etchant, Livingston's dislocation etchant and HCl solution, respectively. Afterward, the etched surfaces were submerged in a solution of tridecafluoroctyltriethoxysilane to modify the composition of the surface. Following the procedure, static water contact angle of 156, 153, and 155 degrees was obtained for aluminum, copper, and zinc surface [7].

The study by Varshney et al.'s [15] prepared a superhydrophobic steel mesh for oil-water separation. The procedure also involved the etching step using a solution of hydrochloric acid and nitric acid and the surface chemical modification with lauric acid. It was subsequently demonstrated that the mesh produced was mechanically, chemically, and thermally stable with static water contact angle of 171 ± 4.5 degrees and sliding angle of 4 ± 0.5 degrees [15].

Lee et al. have implemented a method for reaching a superhydrophobic silicon surface with a water contact angle close to 180 degrees. The roughness of the silicon wafer was increased by immersion in a Cu plating solution, then mixing HF and H_2O_2 to form microstructure and nanostructure. Then, the silicon surface was treated with Teflon for an excellent superhydrophobic surface that can be applied in self-cleaning and microfluidic transport [16].

3.2 Sol-gel method

Sol-gel technique is considered to be the most common method for making superhydrophobic materials, which can be applied on different surfaces of solid substrate surfaces. In this method, colloidal particles of different sizes (from 1 to 100 nm) are dispersed in gels with an inter-connected rigid network with micro-/nano-pores and polymeric chains >1 μm. Here, the monomer is transformed into a colloidal solution (sol) which initiates an integrated network (gel) for polymers or particles.

Several studies on applying this method in superhydrophobic materials have been introduced. Using the sol-gel method, Fan et al. have obtained a superhydrophobic surface on the copper wafer with a water contact angle of 155.4 degrees. After roughening up the copper surface with an acidic solution, a sol-gel coating made of vinyl trimethoxylsilane, ethanol, water, and ammonia was applied. The generated sample can maintain its stability in a 3.5% NaCl solution after the operation, making it potentially useful in anticorrosion and self-cleaning applications [5].

Wang et al. also created superhydrophobic surfaces on a variety of substrates, such as silicon wafers, filter papers, glass slides, electrospun nanofiber mats, and textile fabrics (polyester, wool, and cotton). These surfaces had water contact angles exceeding 170 degrees and sliding angles under 7 degrees. The co-hydrolysis and condensation of tetraethyl orthosilicate (TEOS) and tridecafluorooctyl triethoxysilane (FAS) in NH_3-H_2O-ethanol solution produced the sol solution containing silica nanoparticles. To create a clear film, this sol solution was subsequently deposited onto several substrate surfaces [17].

A technique for producing a superhydrophobic glass surface was reported in the paper by Satapathy et al. A sol-gel SiO_2 nanoparticle solution with a linear low-density polyethylene (LLDPE) polymer matrix was used to coat the glass. Using ethanol as a non-solvent, the porosity of SiO_2 nanoparticles implanted in LLDPE was also altered. The produced SiO_2 nanoparticles embedded in LLDPE matrix, which is exceedingly impressive, had a water contact angle of 170 degrees and a sliding angle of 3.8 degrees.

The sample also had good self-cleaning abilities and was thermally, mechanically, and chemically stable [3].

3.3 Dip-coating method

The substrate is submerged in a component solution that will be placed on the substrate surface when using the dip-coating procedure. The sample is removed from the solution after the specified amount of time, resulting in the film that is then deposited on the substrate. A coating layer forms on the sample surface as a result of the solvent evaporating. This method has a number of benefits, including coating the substrate's upper and lower sides simultaneously, no material waste, application to a variety of materials, high output, and uniform, stable, and long-lasting coating. To create the coating layer, however, all of the components in this process must be submersible.

Mahadik et al. proposed a simple and low-cost dip coating method to fabricate superhydrophobic surface by using organic and inorganic silica precursor methyltrimethoxysilane (MTMS) to produce superhydrophobic silica coatings on the quartz substrate. After the coating step, the recorded water contact angle and water sliding angle were 168 $\pm$ 2 degrees and 3 $\pm$ 1 degrees, respectively. This fabricated surface had remarkable superhydrophobicity and superoleophilicity, along with high durability and optical transparency [18].

In Sun et al.'s study, superhydrophobic surfaces on zinc substrate were fabricated based on Zn's electrochemical processing under an applied electric field. This research applied an electrochemical procedure using an electrolyte mixture consisting of $NaNO_3$ and NaCl to enhance surface roughness, followed by a dip-coating process with fluoroalkylsilane-ethanol solution to modify the surface chemistry of the sample. After two-step process, the zinc surface had a maximum water contact angle of 165.3 degrees and a tilting angle of 2 degrees [19].

Sriram et al. [4] developed a fabrication method for superhydrophobic filter paper using a combination of sol-gel and dip-coating methods. First, the sol-gel solution was prepared by adding poly (methyl methacrylate-co-ethyl acrylate) polymer silicon dioxide nanoparticles to toluene under stirring, followed by adding PFOTS silane for better dispersion of nanoparticles. Then, the filter paper was coated with the solution above via dip-coating method at ambient conditions. The achieved water contact angle of the filter paper surface was >175 degrees, and the sliding angle was 3.8 degrees [4].

3.4 Electrochemical deposition method

The process of electrochemical deposition uses applied voltage to trigger chemical reactions in an aqueous electrolyte solution. This technology has the advantages of being able to apply material in any three-dimensional (3D) geometry and being suited for soft substrates because it is a low-energy process. In addition, it can be carried out at room temperature using water-based electrolytes. The main drawbacks of electrochemical deposition include its limited applicability and insufficient structural strength.

A one-step electrochemical deposition procedure was presented by Huang et al. to create a superhydrophobic surface on a copper substrate. They used a direct voltage (DC) between two copper plates that were submerged in an ethanol-stearic acid solution to carry out the procedure. The anodic copper electrode surface changed to superhydrophobicity as a result of the reaction between copper and the stearic acid solution. SEM pictures showed that the anodic copper surface was covered with copper stearate layers that resembled flowers. The water contact angle on the copper surface was 153 ± 2 degrees with the rolling-off qualities of the droplets [20].

In He and Wang's study [21], this method was utilized to prepare superhydrophobic zinc foil, in which ZnO nanorods were generated on the surface of ZnO film (oxidized zinc foil at 310°C) *via* electrochemical deposition and then spin-coated with perfluoroalkyl methacrylic copolymer to obtain superhydrophobic state. Following the method, a water contact angle of 167 degrees was obtained, demonstrating excellent potential for use in numerous applications, including anti-contamination, anti-fouling, and self-cleaning [21].

Wang et al. presented a three-step process for constructing a biomimetic hierarchical structure on aluminum surface. To create the specimens, anodized porous alumina (APA) film was first made. Nickel and copper were subsequently electrodeposited onto these specimens to create nanometer-sized pillars. To achieve a water contact angle of 152 degrees and a hysteresis angle of 6 degrees for nickel deposited and a water contact angle of 157° and a hysteresis angle of 3 degrees for copper deposited one, the roughed surface was next chemically changed with fluoroalkylsilane [21].

3.5 Plasma-etching method

Plasma-etching involves shooting a suitable gas mixture at a sample with a high stream velocity of glow discharge (plasma). The elements of the etched sample react chemically with the reactive species in the plasma during this process, producing volatile etch products that change the surface of the sample. Plasma treatment could create micro- or nanostructures. The strong points of this technique are the capability of automation, low material consumption, environmental friendliness, and low damage to photoresist. The weak points are high capital investment, difficulty in controlling parameters (surface geometry, types of gases, flow rates, system conductance, patterning, etc.), surface damage caused by plasma radiation, and toxicity of gases.

To create a superhydrophobic zinc surface with a water contact angle of 158 degrees and a sliding angle under 5 degrees, Gao et al. used a two-step plasma-etching procedure. Stearic acid was used to chemically modify the surface after it had been etched with glow discharge electrolysis plasma (GDEP) to increase surface roughness. Finally, the produced zinc surface demonstrated exceptional hydrophobicity, resilience to a range of pH levels, and resistance to prolonged environmental exposure [22].

A straightforward technique for producing superhydrophobic coating on glass was created by Ji et al. using an in-line atmospheric RF glow discharge plasma and a mixture of non-polar aromatic toluene and HMDSO. With a water contact angle of roughly 150 degrees, the fabricated glass demonstrated superhydrophobicity. The employed plasma system could be quickly scaled up for the treatment of larger substrates and continuous processing because it did not require any vacuum equipment and was appropriate for in-line operation [8].

This technique was used in the study by Psarski et al. to create a nanostructured epoxy/Al_2O_3 nanoparticle composite. First, a laser was used to create the microstructures on the metal surface. After that, a plasma source was used to etch the resulting composite in order to add nano-roughness. Once the sample had been etched, it was changed using the dip-coating procedure with 1H,1H,2H,2H perfluorotetradecyl-triethoxysilane to obtain a superhydrophobic state with a contact angle of 160 degrees and a sliding angle of 8 degrees [23].

3.6 Hydrothermal method

The hydrothermal process results in the formation of crystalline materials in a hot aqueous solution under high vapor pressure. By applying high pressure and temperature to the sample surface, this technique is typically employed to produce roughness. Crystals are produced in an autoclave during the procedure using the substrate and water that is given. Because to its simplicity, this process can produce huge crystals of excellent quality and is suitable for commercial use. The apparatus cost for this method is also significant, and the crystal formation cannot be observed or controlled during the procedure.

Shi et al. suggested using a hydrothermal process to create superhydrophobic glass. In order to increase surface roughness, the glass sample was hydrothermally treated with a solution of $Al(NO_3)_3 \cdot 9H_2O$, NaOH, and colloidal silica. This was followed by chemical modification using octyltrimethoxysilane to obtain superhydrophobicity state. Following the procedure, the prepared sample obtained a 154 degrees water contact angle and a sliding angle that was <3 degrees [11].

Wu et al. suggested using an alkaline hydrothermal process to create superhydrophobic surfaces with a ZnO micro- and nanostructure. Rose-like structures were present on the surface of the treated sample following the hydrothermal stage, which increased surface roughness. After the subsequent process, which involved spin-coating the sample with Teflon, it attained superhydrophobicity with a contact angle of 168 degrees and good stability (no change after being immersed in water for 15 days). In general, the suggested approach is easy to use, affordable, and possibly useful in a wide range of sectors [24].

The hydrothermal approach on a microstructured Ti_6Al_4V alloy surface was one of three procedures used in Shen et al. to create various nanostructures. A two-step chemical reaction, anodic oxidation, and the hydrothermal technique were used to make nanowire, nanotube, and nanomesh structures. In order to obtain the surface's extraordinary superhydrophobicity, FAS-17 was used to chemically modify it. This resulted in a water contact angle of 161 degrees and a tilting angle of 6 degrees [25].

3.7 Self-assembly method

In the self-assembly process, a disordered system of preexisting components assembles in an organized manner or a sequence thanks to interactions of local non-covalent molecules. With minimal interference, complex structures can be formed. The layer can grow using this technology at low temperatures without the need for expensive machinery or involved procedures. This type of surface treatment technique is therefore extensively applicable. Unfortunately, this procedure has a complex mechanism and takes a very long period.

Superhydrophobic copper was created by Yin et al. [26] using chemical etching and self-assembly techniques. In order to produce flower-like structures and nanoneedles of $Cu(OH)_2$ on the sample surface, the copper was first etched with a solution of sodium hydroxide and potassium persulfate. Finally, a self-assembled surface layer was created by submerging the roughed-up sample in dodecanoic acid.

When everything was finished, the prepared surface had a good superhydrophobic state with a 153 degrees WCA [26].

In the study by Song et al. [27], a technique for creating silicon with a superhydrophobic surface was presented. The aluminum-induced crystallization of the amorphous silicon technology produced the micro/nano textures on the silicon substrate surface. The next step was putting an octadecyltrichlorosilane self-assembled monolayer on silicon textured surface. The manufactured surfaces attained a 155 degrees water contact angle and a <1 degree slide angle [27].

Table 1 summarized the fabrication technique for superhydrophobic materials and their applications.

4. Applications of superhydrophobic materials in environmental treatment

Due to the unique chemical compositions and characteristics, many pollutants that are now present in water are incredibly challenging to remove. To address this problem, which has an immediate impact on both the environment and human health, scientists and researchers have put in a tremendous amount of work. The superhydrophobic surface is covered in this book chapter as a cutting-edge material that can be used in many methods of combating environmental pollution.

4.1 Oil removal

In recent years, oil spills and microplastic have become vital worldwide environmental issues. Besides different treatment methods like combustion, chemical treatment, or bioremediation, which may cause secondary pollution after being applied, the utilization of superhydrophobic material appears promising. This method can help separate the oil from the water, returning the polluted water to its original state without generating secondary pollution after pass through the superhydrophobic materials (**Figure 3**). For instance, the combustion method can help remove the oil from the water, but it also causes air pollution because it produces a large amount of CO_2 and SO_2 after the procedure.

Several published studies developed superhydrophobic surfaces and applied them in oil-water separation. For instance, Yeom and Kim proposed a dip-coating method using silica nanoparticles, hexadecyltrimethoxysilane to fabricate a superhydrophobic surface on steel mesh and sponge. The obtained water contact angles were 151.9 ± 1.6 degrees and 152.4 ± 3.2 degrees for steel mesh and sponge, respectively. Both fabricated surfaces then showed excellent performance in oil–water separation [28].

Crude oil and kerosene were often used as oil components in oil-water separation because of their close relation with oil spill pollutants, which can lead to disastrous consequences for the environment and society. Xue et al. introduced a sol-gel method with tetraethoxysilane and 1,1,1,3,3,3-hexamethyl disilazane as precursors for preparing superhydrophobic textiles. After the fabrication, the fabricated textile possessed superhydrophobic and superoleophilic properties, and it could successfully separate the crude oil from the water in an oil–water mixture [29].

Yin et al. proposed a hydrothermal method using nickel sulfide, thioacetamide, sodium hydroxide cetyltrimethylammonium bromide to fabricate superhydrophobic nickel mesh for oil-water separation purposes. The achieved WCA after modification was 158 degrees, indicating good superhydrophobicity. The fabricated mesh also

No	Substrate	Fabrication method	Properties	Application	Ref
1	Steel mesh-based material modififed with HNO_3/HCl and lauric acid	Chemical etching method	• Water contact angle: 171 ± 4.5 degrees • Sliding angle: 4 ± 0.5 degrees	Oil-water separation	[15]
2	Silicon substrate modififed with HF/H_2O_2	Chemical etching method and spin-coating	• Water contact angle: nearly 180 degrees	Structured channel for the fast transport of microfluidics and self-cleaning surfaces	[16]
3	Copper wafers modififed with vinyltrimethoxylsilane (VTMS)	Sol-gel method	• Water contact angle: 155.4 degrees	Resistance to corrosion and selfcleaning application	[5]
4	Microscope glass slides modififed with linear low density polyethylene (LLDPE) and silicon dioxide nanopowder	Sol-gel method	• Water contact angle: 170 degrees • Sliding angle: 3.8 degrees	Packaging and other industrial applications.	[3]
5	Cleaned quartz substrates modififed with methyltrimethoxysilane (MTMS), CH_3OH, and oxalic acid	Dip-coating method	• Water contact angle: 168 ± 2 degrees • Sliding angle: 3 ± 1 degrees	Material against outdoor environmental conditions	[18]
6	Filter paper modififed with poly (methyl methacrylate-coethyl acrylate) (PMMA), 1H,1H,2H,2H Perfluorooctyltrichlorosilane (PFOTS)	Sol-gel and dip-coating methods	• Water contact angle: 175 degrees • Sliding angle: 3.8 degrees	Oil-water separation	[4]
7	Copper surface	Electrochemical deposition method	• Water contact angle: 153 ± 2 degrees	Corrosion resistance application	[20]
8	ZnO film	Electrochemical deposition method	• Water contact angle: 167 degrees	Anti-contamination, anti-fouling, and self-cleaning	[21]
9	Zinc surface	Plasma-etching method	• Water contact angle: 158 degrees • Sliding angle: 5 degrees	Corrosion resistance application	[22]
10	Nanocomposite replicas of aluminum modified with 1H,1H,2H,2H perfluorotetradecyltriethoxysilane	Plasma-etching method and dip-coating	• Water contact angle: 160 degrees • Sliding angle: 8 degrees	Corrosion resistance for industrial applications	[23]
11	Glass substrate kept at 160°C and modified with octyltrimethoxysilane	Hydrothermal method	• Water contact angle: 154 degrees • Sliding angle: 3 degrees	Self-cleaning applications	[11]
12	ZnO surface kept at 180°C	Hydrothermal method	• Water contact angle: 168 degrees	Industrial applications	[24]
13	Copper plates modified with sodium hydroxide and potassium persulfate	Chemical etching and Self-assembly method	• Water contact angle: 153 degrees	Engineering metal applications	[26]
14	Silicon wafers with sodium octadecyltrichlorosilane	Self-assembly method	• Water contact angle: 155 degrees • Sliding angle: 1 degrees	Electro-mechanical system applications	[27]

Table 1.
Fabrication technique for superhydrophobic materials and their applications.

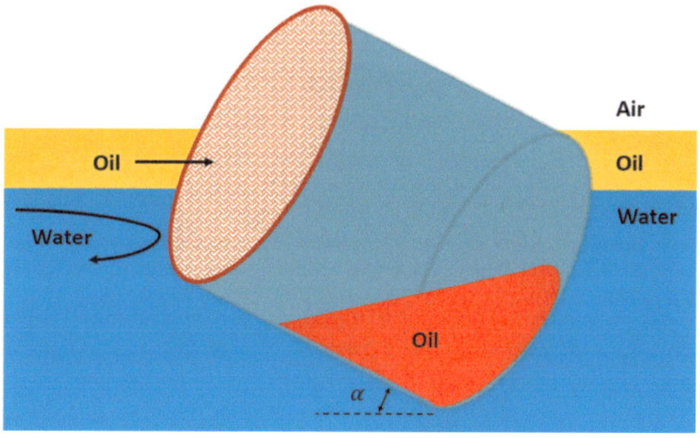

Figure 3.
Example on oil-water separation method.

showed exemplary implementation in the separation experiment of kerosene-water mixture, with an average recovery rate of over 95% [30].

Moreover, filter paper was chosen as the production substrate for superhydrophobic surfaces in other studies. Filter paper has various benefits over other substrates, including biodegradability, high flexibility, and affordability. In the study by Zhang et al., TEOS, octadecyltrichlorosilane, and polystyrene were used to create a superhydrophobic surface on filter paper. After the treatment, the paper surface's wettability changed from being superhydrophilic to being superhydrophobic, with a 156 degrees WCA. Hexan-water separation was another area where the manufactured paper excelled [31].

Teng et al. developed a coating method with nano $TiO2$, γ-aminopropyl-triethoxysilane, and polydimethylsiloxane (PDMS) to prepare superhydrophobic filter paper. The resulting water contact and rolling angles were 154.5 and 3.5 degrees, respectively. The prepared paper also exhibited good anti-fouling, self-cleaning, and oil-water separation ability, which was evaluated through a filtration experiment of dichloromethane-water mixture [32].

However, many hazardous chemicals are being used in these studies, such as hexadecyltrimethoxysilane, aminopropyltriethoxysilane, and octadecyltri-chlorosilane. Besides, most procedures require more than a day to fabricate superhydrophobic surface, which can be considered time-consuming. Some studies have yet to explicitly report the fabricated surface's durability and reusability. Thus, in this study, a fast and environmental-friendly method for the fabrication of superhydrophobic surface on filter paper will be developed. In addition, the applicability in oil-water separation (kerosene as oil component), durability, and reusability will also be examined after the fabrication procedure to evaluate the full potential of the fabricated filter papers.

4.2 Desalination

A useful, eco-friendly, and practical solution to the freshwater problem is solar desalination *via* interfacial evaporation. It is still difficult to create high-quality light absorber materials that can operate in a variety of tough environments and minimize heat loss while avoiding salt buildup during seawater evaporation. Due to their ability to prevent water from penetrating the pores of the absorber, superhydrophobic surfaces can be employed to circumvent this problem and open up space for vapor to pass through.

Azeem et al. created a superhydrophobic polydimethylsiloxane (PDMS) membrane using an easy and affordable technique. A surface with textures like the lotus effect can be seen on a PDMS-based membrane, which is interesting. PDMS-based textured membrane demonstrated excellent MD performance with high and sustained salt rejection at 99.99%, and permeate flux exceeding 21 kg m^2 h during a 40-hour test using very saline water (70 g/l NaCl) in air-gap membrane distillation (AGMD). Significantly, the PVDF membranes produced in this work demonstrated higher permeate flux over time as a result of a narrowing of the air gap caused by membrane stretching [33].

Superhydrophobic membranes are crucial for enhanced seawater desalination, according to a study by Ray et al. By straightforward sol-gel processing, they also achieved the successful fabrication of a three-layered membrane with a top superhydrophobic coating onto a polypropylene (PP) mat. With this membrane, a high permeate flux of roughly 6.7 liters per square meter per hour (LMH) was maintained for 16 hours, and the salt rejection level was estimated to be 99.7% [34].

An electrospun nanofibrous membrane with superhydrophobicity was used by Zhou et al. as a promising candidate for membrane distillation. With a static contact angle of 156.6 ± 1.38 degrees and a small sliding angle of 6.4 ± 0.2 degrees, the ideal membrane demonstrates superhydrophobicity. The best membrane in a laboratory-scale direct contact membrane distillation has a high permeate flux of 38.8 kg m^2 h and a salt rejection of 99.99% during the course of a 40-hour long desalination operation. With regard to wetting and fouling resistance, the superhydrophobic membrane excels.

4.3 Microplastic removal

As the world's capacity to deal with the rapidly rising output of disposable plastic goods exceeds it, plastic pollution has also emerged as one of the most urgent environmental challenges. In impoverished Asian and African countries, where rubbish collection services are either ineffective or nonexistent, plastic pollution is most noticeable. According to the report of the Environmental program of the United Nations, plastic pollution is a global problem when approximately 7 billion of the 9.2 billion tons of plastic produced from 1950 to 2017 became plastic waste, ending up in landfills or dumped [35]. Seriously, most plastics are difficult to degrade and can persist in ecological environments for many years, and they could accumulate in animal bodies and cause long-term damage. Several papers mentioning plastic (including microplastic, microplastic, and nanoplastic) removal have been published recently, showing humans' extremely high seriousness and interest in this environmental problem. These microplastics have been detected in many places, such as on land, sea, sediments, polar regions, the atmosphere, and drinking water. In the aquatic environment, the surface of microplastics is easily changed. As they get rougher and more negatively charged, they can effectively serve as a mobile surface to disperse a variety of organic contaminants, including animals and dangerous microbes. Microplastics in the environment are primarily produced at wastewater treatment facilities. But none of the water treatment techniques used today are specifically developed to get rid of microplastics. A certain quantity of microplastics can still be released into the water and continue to accumulate even though wastewater treatment plants can remove the majority of them (>90%). It is estimated that wastewater

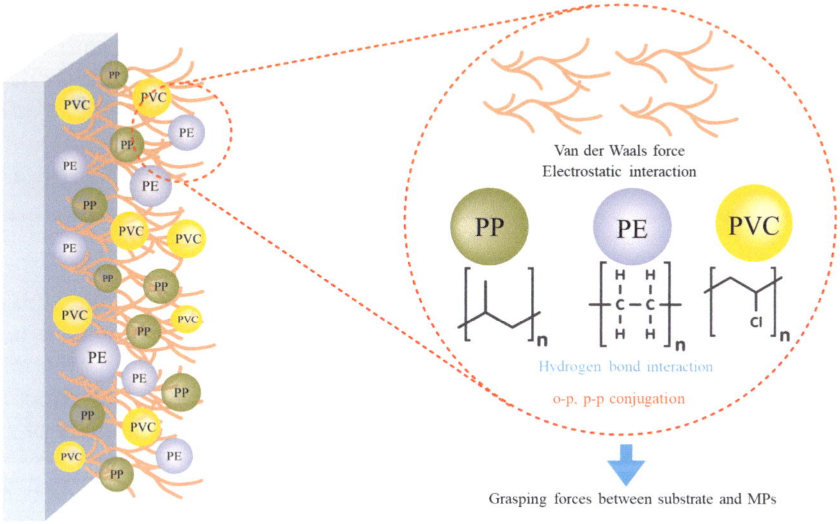

Figure 4.
Microplastics removal by adsorption process.

treatment facilities discharge between 15,000 and 4.5 million microplastic particles into surface water each day [36].

In order to provide a quick and effective method of eliminating microplastics from the environment, numerous approaches and materials have been developed. For instance, the flocculation-flocculation method may get rid of between 75.6% and 85.2% of microplastics [37]. Moreover, membrane bioreactors may remove over 95% microplastic (more than 20 m), but they are expensive to operate and membrane fouling is a possibility [38]. For the elimination of microplastics, the adsorption approach has received substantial study as an efficient, affordable, and user-friendly technique. The polluted particles can be kept in the surface of the superhydrophobic materials by the electrostatic interaction, hydrogen bonding with the microplastic particles (**Figure 4**). Electrostatic interaction, hydrogen bonding, and interaction are the main components of the adsorption process employed by Sun et al to remove polystyrene microplastics from porous materials based on chitin [39]. Several research have also suggested superhydrophobic surfaces, which can be used to adsorb microplastics from aqueous NaCl with an efficiency higher than 99%. Although there has been little study of water in natural settings and less attention paid to environmental durability, the majority of research has concentrated on simulating water trials in the lab.

Ayra et al. reported on the creation of a superhydrophobic 304 stainless steel mesh using chemical etching and lauric acid liquid-phase deposition. A 304 stainless steel mesh's surface was altered by oxidizing conditions ($FeCl_3/HCl/H_2O_2$) and liquid-phase deposition of lauric acid, which resulted in the surface achieving superhydrophobicity (169 degrees) and superoleophilicity (0 degrees). Using the superoleophilic qualities of the modified 304 SS mesh, this mesh was used to remove HDPE microplastic that HDPE migrated from the aqueous phase to the organic phase. It is explained that the repellent cloud that forms on the surface of the microplastic in water enhances dispersion. Yet, in the organic phase, van der Waals interactions cause microplastic to aggregate [40].

According to Ayra et al., superhydrophobic surfaces and functionalized surfaces for microplastics should be taken into account when choosing between different processes and methods for removing microplastics. The solid contaminants are given hydrophilic characteristics by microplastic surface functionalization (with ClO or Fe_3O_4), enhancing selectivity in the separation process. While the oil used affects selectivity when employing superhydrophobic surfaces, microplastics can also be distinguished based on their chemical makeup. The creation of safer goods and removal methods for solid contaminants will be a major problem in the future [41].

4.4 Air filter

The use of superhydrophobic nonwovens is progressively expanding, from the initial oil-water separation and self-cleaning to air filtration, waterproof self-healing, antibacterial, and other disciplines, with the advancement of society and the raising of people's standards of life. Superhydrophobicity can improve the barrier effect of masks on dust particles in the field of air filtration, particularly in the field of masks. Respiratory droplets are used to disseminate the highly infectious coronavirus. Virus-carrying droplets can adhere to the surface of modern surgical masks while being worn, placing users at risk of infection. Superhydrophobic nonwoven masks can, however, prevent the majority of droplets from sticking to their surface, boosting their ability to act as a barrier against droplets, and lowering the risk of COVID-19 infection.

A superhydrophobic, photo-sterilized, and reusable mask that may be worn for an extended period of time and used again after solar illumination was studied by Lin et al. The mask was created by ultrasonicextrusion and is based on graphene nanosheet-embedded carbon (GNEC) layer to protect COVID-19. Excellent characteristics of the GNEC mask include being superhydrophobic (water contact angle: 157.9 degrees), having a 100% BFE, and being photo-sterilizing (photothermal performance: 110.6°C). This work might encourage individuals to investigate surgical mask performance improvements to sustain global health and development [42].

Both the barrier effect of droplets and the barrier effect of particles can be improved by adding superhydrophobic characteristics. The introduction of nanoparticles into filtering nonwovens is being driven by these benefits for many researchers.

4.5 Self-healing

Polymers, metals, ceramics, and their composites are examples of self-healing materials because they can either fully or partially regain their original set of properties after being harmed during usage. Superhydrophobic nonwovens have been increasingly added to self-healing in recent years by researchers. By continuously producing the epicuticle wax layer or by allowing the micro- or nanostructure to naturally regenerate after being damaged, plant leaves, bird feathers, or insect wings can effectively preserve their ability to repel liquids in contrast to manufactured structures. Researchers have created several self-healing superhydrophobic surfaces to increase durability and extend the life of outdoor applications, drawing inspiration from the self-healing characteristics of natural plants and animals. In example, the superhydrophobic surfaces of sponge can be easily recovered after absorbing the pollutants in the aqueous solution (**Figure** 5).

Wang and colleagues imitated the Lotus leaf's structure. Using the use of an easy replica molding technique, they created a biomimetic way to create a self-healing superhydrophobic surface by integrating n-nonadecane wax into a microstructured

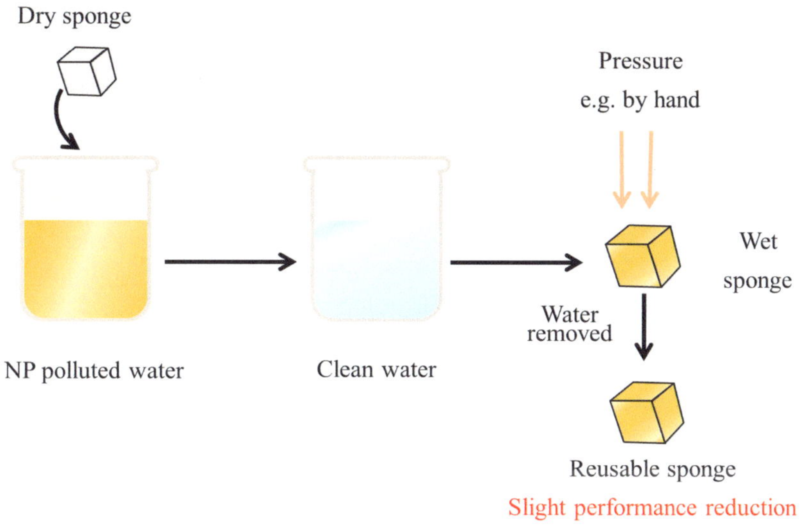

Figure 5.
Self-healing property of sponge.

PDMS matrix. Without any external stimulation, the injured surface might swiftly regain its superhydrophobicity [43].

Superhydrophobic surfaces with built-in liquid repellent qualities have a wide range of potential applications. In the study by Li et al., thermoplastic polyurethane PU/carbon nanotube (CNT) composite films were created by filtering CNT suspensions *via* electrospun thermoplastic polyurethane nonwovens. The strain-gauge sensors displayed outstanding self-restoring water repellency, high sensitivity, and durability [44].

4.6 Antibacterial

Bacterial adhesion causes financial loss, medical difficulties, and even fatalities, making it a major worry in the medical community globally. Due to the rapid emergence of bacterial resistance to the overuse and abuse of antibiotics, new approaches to combatting bacterial infections are urgently needed. Growing interest has been shown in bioinspired superhydrophobic coatings as a novel method of preventing bacterial adhesion and illnesses. In example, by depositing copper and stearic acid on cotton fabric, Suryaprabha et al. effectively created a straightforward, self-cleaning, sturdy, chemically resilient, and cost-effective antibacterial superhydrophobic coating. Gram-positive and Gram-negative bacteria were used to test the antibacterial activity of the created coating, and the inhibition zone method showed that the cotton fabric's ability to resist blood stains [45].

5. Conclusions and outlook

It is interesting to note that the quantity of papers on new superhydrophobic materials has largely held steady since 2019. Yet, the number of published works on the use of superhydrophobic materials in the environment is still constantly increasing. It demonstrates interest in the topic. Due of their numerous possible uses, biomimetic superhydrophobic materials have garnered a lot of attention during the past 10 years. The creation of superhydrophobic materials is no longer surprising given recent technical advancements. Yet, it is still difficult to create anti-corrosive coatings that are strong, resilient, affordable, and ecologically benign, and this field is still young. There are still a lot of issues that need to be solved, as well as some shortcomings and difficulties in industrial manufacturing and practical applications. Further research on superhydrophobic surfaces should concentrate on enhancing their stability and toughness. Further research is needed to better understand how to precisely manage surface structure and roughness on both microscopic and large scales. In addition, it is best to avoid using halogenated polymers or fluorine- and chlorine-containing compounds when making superhydrophobic materials. Superhydrophobic materials still have a lot of great uses in the environment, despite the fact that all the issues described above need to be resolved.

Acknowledgements

This work is fully supported by the project with the code number of VJU. JICA.21.03, from VNU Vietnam Japan University, under Research Grant Program of Japan International Cooperation Agency (JICA).

Conflict of interest

The authors declare no conflict of interest.

Author details

Thi Viet Ha Tran[1*], Minh Viet Nguyen[2] and Le Minh Tri Nguyen[3]

1 Faculty of Advanced Technology and Engineering, VNU Vietnam – Japan University, Ha Noi, Viet Nam

2 VNU Key Laboratory of Advanced Material for Green Growth, Faculty of Chemistry, VNU University of Science, Hanoi, Viet Nam

3 Laboratory of Advanced Materials Chemistry, Advanced Institute of Materials Science, Ton Duc Thang University, Ho Chi Minh City, Vietnam

*Address all correspondence to: ttv.ha@vju.ac.vn

References

[1] Das S, Kumar S, Samal SK, Mohanty S, Nayak SK. A review on superhydrophobic polymer nanocoatings: recent development and applications. Industrial & Engineering Chemistry Research. 2018;**57**(8): 2727-2745

[2] Chevallier P, Turgeon S, Sarra-Bournet C, Turcotte R, Laroche G. Characterization of multilayer anti-fog coatings. ACS Applied Materials & Interfaces. 2011;**3**:750-758

[3] Satapathy M, Varshney P, Nanda D, Mohapatra SS, Behera A, Kumar A. Fabrication of durable porous and non-porous superhydrophobic LLDPE/ SiO_2 nanoparticles coatings with excellent self-cleaning property. Surface and Coatings Technology. 2018;**341**: 31-39

[4] Sriram S, Singh RK, Kumar A. Oil-water separation through an ultrahydrophobic filter paper developed by sol-gel dip-coating technique. Materialstoday: Proceedings. 2020;**26**: 2495-2501

[5] Fan Y, Li C, Chen Z, Chen H. Study on fabrication of the superhydrophobic sol–gel films based on copper wafer and its anti-corrosive properties. Applied Surface Science. 2012;**258**(17):6531-6536

[6] Song Y, Nair RP, Zou M, Wang Y. Superhydrophobic surfaces produced by applying a self-assembled monolayer to silicon micro/nano-textured surfaces. Nano Research. 2009;**2**:143-150

[7] Qian B, Shen Z. Fabrication of superhydrophobic surfaces by dislocation-selective chemical etching on aluminum, copper, and zinc substrates. Langmuir. 2005;**21**(20):9007-9009

[8] Ji YY, Kim S-S, Kwon O-P, Lee S-H. Easy fabrication of large-size superhydrophobic surfaces by atmospheric pressure plasma polymerization with non-polar aromatic hydrocarbon in an in-line process. Applied Surface Science. 2009;**255**(8): 4575-4578

[9] Rezaei S, Manouchehri I, Moradian R, Pourabbas B. One-step chemical vapor deposition and modification of silica nanoparticles at the lowest possible temperature and superhydrophobic surface fabrication. Chemical Engineering Journal. 2014;**252**:11-16

[10] Ge B, Li Y, Han B, Li H, Ren G, Li W, et al. Carrier lifetime regulation strategy to improve the sewage purification capacity of superhydrophobic silver phosphate fabric. Advanced Powder Technology. 2020;**31**(12):4691-4698

[11] Shi F, Chen X, Wang L, Niu J, Jihong Y, Wang Z, et al. Roselike microstructures formed by direct in situ hydrothermal synthesis: from superhydrophilicity to superhydrophobicity. Chemistry of Materials. 2005;**17**:6177-6180

[12] Movahedi T, Norouzbeigi R. Synthesis of flower-like micro/nano ZnO superhydrophobic surfaces: additive effect optimization via designed experiments. Journal of Alloys and Compounds. 2019;**795**:483-492

[13] Sethi SK, Manik G, Sahoo SK. Chapter 1 - Fundamentals of superhydrophobic surfaces. In: Samal SK, Mohanty S, Nayak SK, editors. Superhydrophobic Polymer Coatings. Elsevier; 2019. pp. 3-29. DOI: 10.1016/ B978-0-12-816671-0.00001-1. ISBN

9780128166710. Available from: https://www.sciencedirect.com/science/article/pii/B9780128166710000011

[14] Young T. An essay on the cohesion of fluids. Philosophical Transactions of the Royal Society of London. 1805;**95**:65-87

[15] Varshney P, Nanda D, Satapathy M, Mohapatra SS, Kumar A. A facile modification of steel mesh for oil–water separation. New Journal of Chemistry. 2017;**41**:7463-7471

[16] Lee J-P, Choi S, Park S. Extremely superhydrophobic surfaces with micro- and nanostructures fabricated by copper catalytic etching. Langmuir. 2011;**27**(2): 809-814

[17] Wang H, Fang J, Cheng T, Ding J, Qu L, Dai L, et al. One-step coating of fluoro-containing silicananoparticles for universal generation of surface superhydrophobicity. Chemical Communications. 2008;**2008**:877-879

[18] Mahadik SA, Parale V, Vhatkara RS, Mahadik DB, Kavale MS, Wagh PB, et al. Superhydrophobic silica coating by dip coating method. Applied Surface Science. 2013;**277**:67-72

[19] Sun J, Zhang F, Song J, Wang L, Qu Q, Lu Y, et al. Electrochemical fabrication of superhydrophobic Zn surfaces. Applied Surface Science. 2014; **315**:346-352

[20] Huang Y, Sarkar DK, Chen X-G. A one-step process to engineer superhydrophobic copper surfaces. Materials Letters. 2010;**64**(24): 2722-2724

[21] He G, Wang K. The super hydrophobicity of ZnO nanorods fabricated by electrochemical deposition method. Applied Surface Science. 2011; **15**:6590-6594

[22] Gao J, Li Y, Li Y, Liu H, Yang W. Fabrication of superhydrophobic surface of stearic acid grafted zinc by using an aqueous plasma etching technique. Open Chemistry. 2012;**10**(6):1766-1772

[23] Psarski M, Marczak J, Grobelny J, Celichowski G. Superhydrophobic surface by replication of laser micromachined pattern in epoxy/alumina nanoparticle composite. Journal of Nanomaterials. 2014;**2014**. Article ID 547895. DOI: 10.1155/2014/547895

[24] Wu J, Xia J, Lei W, Wang B-P. Fabrication of superhydrophobic surfaces with double-scale roughness. Materials Letters. 2010;**64**(11):1251-1253

[25] Shen Y, Tao H, Chen S, Xie Y, Zhou T, Wang T, et al. Water repellency of hierarchical superhydrophobic Ti6Al4V surfaces improved by secondary nanostructures. Applied Surface Science. 2014;**321**:469-474

[26] Yin S, Wu D, Yang J, Lei S, Kuang T, Zhu B. Fabrication and surface characterization of biomimic superhydrophobic copper surface by solution-immersion and self-assembly. Applied Surface Science. 2011;**257**(20): 8481-8485

[27] Song Y, Nair RP, Zou M, Wang Y. Superhydrophobic surfaces produced by applying a self-assembled monolayer to silicon micro/nano-textured surface. Nano Research. 2009;**2**:143-150

[28] Yeom C, Kim Y. Purification of oily seawater/wastewater using superhydrophobic nano-silica coated mesh and sponge. Journal of Industrial and Engineering Chemistry. 2016;**40**: 47-53

[29] Xue C-H, Ji P-T, Zhang P, Li Y-R, Jia S-T. Fabrication of superhydrophobic and superoleophilic textiles for oil–water

separation. Applied Surface Science. 2013;**284**:464-471

[30] Yin X, Yu S, Wang L, Li H, Xiong W. Design and preparation of superhydrophobic NiS nanorods on Ni mesh for oil-water separation. Separation and Purification Technology. 2020;**234**:116126

[31] Zhang M, Wang C, Wang S, Shi Y, Li J. Fabrication of coral-like superhydrophobic coating on filter paper for water–oil separation. Applied Surface Science. 2012;**261**:764-769

[32] Teng Y, Wang Y, Shi B, Li X, Chen Y. Robust superhydrophobic surface fabrication by fluorine-free method on filter paper for oil/water separation. Polymer Testing. 2020;**91**:106810

[33] Azeem MA, Lawal DU, Abdulgader HA, Baroud TN. Enhanced performance of superhydrophobic polyvinylidene fluoride membrane with sandpaper texture for highly saline water desalination in air-gap membrane distillation. Desalination. 2022;**528**:115603

[34] Ray SS, Chen S-S, Chang H-M, Thanh CND, Le HQ, Nguyen NC. Enhanced desalination using a three-layer OTMS based superhydrophobic membrane for a membrane distillation process. RSC Advances. 2018;**8**(18): 9640-9650

[35] EpoU N. Plastic Pollution 2022. Available from: https://www.unep.org/plastic-pollution

[36] Yang L, Li K, Cui S, Kang Y, An L, Lei K. Removal of microplastics in municipal sewage from China's largest water reclamation plant. Water Research. 2019;**155**:175-181

[37] Iyare PU, Ouki SK, Bond T. Microplastics removal in wastewater treatment plants: a critical review. Environmental Science: Water Research & Technology. 2020;**6**: 2664-2675

[38] Talvitie J, Mikola A, Koistinen A, Setälä O. Solutions to microplastic pollution – removal of microplastics from wastewater effluent with advanced wastewater treatment technologies. Water Research. 2017;**123**:401-407

[39] Sun C, Wang Z, Zheng H, Chen L, Li F. Biodegradable and reusable sponge materials made from chitin for efficient removal of microplastics. Journal of Hazardous Materials. 2021;**420**:126599

[40] Rius-Ayra O, Biserova-Tahchieva A, Sansa-López V, Llorca-Isern N. Superhydrophobic 304 stainless steel mesh for the removal of high-density polyethylene microplastics. Langmuir. 2022;**38**(18):5943-5953

[41] Rius-Ayra O, Biserova-Tahchieva A, Llorca-Isern N. Surface-functionalised materials for microplastic removal. Marine Pollution Bulletin. 2021;**167**:112335

[42] Lin Z, Wang Z, Zhang X, Diao D. Superhydrophobic, photo-sterilize, and reusable mask based on graphene nanosheet-embedded carbon (GNEC) film. Nano Research. 2021;**14**(4): 1110-1115

[43] Wang Y, Liu Y, Li J, Chen L, Huang S, Tian X. Fast self-healing superhydrophobic surfaces enabled by biomimetic wax regeneration. Chemical Engineering Journal. 2020;**390**:124311

[44] Li Z, Ye L, Shen J, Xie K, Li Y. Strain-gauge sensoring composite films with self-restoring water-repellent properties for monitoring human movements. Composites Communications. 2018;**7**:23-29

[45] Suryaprabha T, Sethuraman MG. Fabrication of copper-based superhydrophobic self-cleaning antibacterial coating over cotton fabric. Cellulose. 2017;**24**:395-407

Chapter 2

Plant-Based Sustainable Self-Cleaners in Nanotechnology Era: From Mechanism to Assembling

Mojdeh Nakhaei, Jing Ying Chong, Yunlong Tang and Shahnaz Mansouri

Abstract

Nature has always been a resource of inspiration for humans, providing valuable lessons that have led to innovative solutions throughout history. Observing the micro-nano roughness structure of bio-surfaces has led to the discovery of natural self-cleaning surfaces for over 25 years. This has sparked a new field of research with valuable applications. Numerous self-cleaning products made from plant extracts have been created by replicating the natural purifying abilities of plant surfaces. Significant literature exists on the development, classification, extraction, and production of self-cleaning agents for diverse industries through a thorough understanding of bio-cleaning mechanisms. Various methods have been developed to synthesize these surfaces, including immersion, electrochemical deposition, emulsion, electrospinning, phase-separation, Chemical-Vapor-Deposition (CVD), spray coating, wet chemical reaction, and three-dimensional printing (3D-printing), among others. Currently, the primary objective is to gain knowledge from nature and utilize it to develop novel products for food, pharmaceutical, and related industries. Natural plant-based self-cleaning surfaces can be characterized by their superhydrophobicity and superhydrophilicity regimes. The process of 3D-printing is a computer-based technique that builds up three-dimensional objects through the layer-by-layer deposition of materials. The creation of effective self-cleaning surfaces with unique wettability, chemical properties, and microstructure depends on the design and engineering of solid surfaces.

Keywords: plant-based, self-nano-cleaners, mechanism, food packaging, wettability, 3D printing, hydrophilic, hydrophobic

1. Introduction

In 1997, Barthlott and Neinhuis [1] discovered that the micro- and nano-roughness of lotus leaves accounted for the self-cleaning phenomenon. A self-cleaning surface maintains its cleanliness through the action of water, which lowers maintenance costs, requires less work, and requires fewer detergents [2]. By understanding this natural phenomenon, scientists started to mimic the structural properties of lotus leaves and apply them to different products, including self-cleaning roof tiles and paints and so many others. The main concept is based on two wettability regions hydrophilic (drawn to water/ water-attractant) and hydrophobic (away from water/water-repellent) phenomena.

This chapter will introduce the latest research developments in bioinspired self-cleaning surfaces inspired by plants. Specifically, we will examine a variety of plant-based self-nano-cleaners, investigating their mechanisms and exploring the diverse applications they have already been utilized in across various industries. Three-dimensional printing (3D-printing) is discussed as a new developing process. The process of creating objects through layer-by-layer computer design and building up materials is known as 3D-printing, also called additive manufacturing. This innovative technique allows for the creation of three-dimensional pieces.

2. Plant-based self-nano-cleaners in nature

Self-cleaning technology has made significant advancements since the late 20th century, resulting in numerous valuable applications. This technology has a wide range of uses, from solar cell panels to window glass.

In order to develop self-cleaning surfaces, it is important to identify the roles, structures, and underlying bases of different objects in nature that exhibit self-cleaning properties. Scientists have gained valuable insights from studying living nature, which has led to the creation of highly effective bioinspired surfaces. Water is an ideal medium for removing various types of contaminants from surfaces due to its abundance and ideal density and polarity. Nature has provided many innovative designs for using water energies and surface properties to clean material surfaces [2, 3].

Superhydrophobicity and superhydrophilicity are two distinct wettability regimes that apply to plant-based self-cleaning surfaces in nature. **Table 1** shows a number of plants that are introduced as plant-based self-cleaning surfaces in literature.

Surface	Plant (leaves)	wettability regimes [Reference]
Hydrophobic	Lotus	superhydrophobic, low adhesion, low drag [2–6]
	Rice	superhydrophobic, Anisotropic wetting [2, 3, 7, 8]
	India canna	superhydrophobic [2, 3]
	Taro	superhydrophobic [2, 3]
	Cabbage	superhydrophobic [9]
	Indian cress	superhydrophobic [9]
	Salvinia molesta	superhydrophobic hairs [10]

Surface	Plant (leaves)	wettability regimes [Reference]
Hydrophilic	Pitcher	superhydrophilic, low drag (slippery) [2]
	Anubias barteri	superhydrophilic [2, 3, 6]
	Heliamphora nutans	superhydrophilic [2, 3, 6]

Table 1.
Examples of plant-based self-cleaning surfaces.

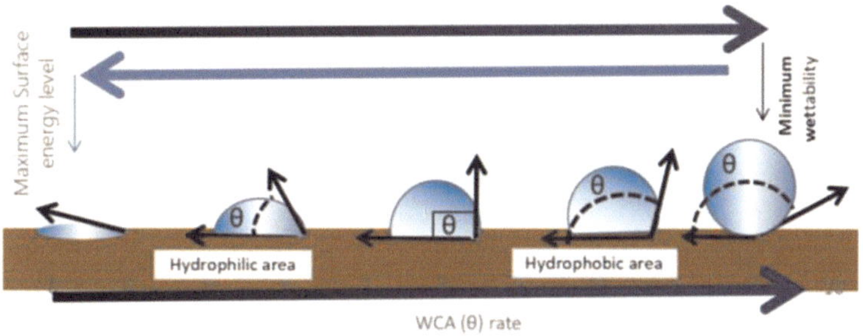

Figure 1.
Liquid drop schematic to show water contact angle (WCA) based on surface energy level ~ surface wettability.

Chemical characteristics and surface microstructure of a solid surface affect its wettability. If a surface has a water-contact-angle (WCA) greater than 90° is well-defined as a hydrophobic surface, and within WCA, more than 150°is a superhydrophobic surface. These surfaces are water-repellent. If a surface has a WCA lower than 90°, it is identified as a hydrophilic surface, and by displaying, WCA lower than 10° is a superhydrophilic surface. Superhydrophilic surfaces are extremely wettable and can reduce water contact angles (close to zero) to create water films (**Figure 1**) [2, 3].

3. Mechanisms of these self-control-cleaners

Plant surfaces have a variety of wettability characteristics, from wax-free, wet floating leaves to water-repellent leaves growing on land. The leaves of some plants are arguably well-known representatives of water repellency on the planet (**Table 1**). Bacterial and fungal spores are among the pathogens found in free water, which is protected by water-repellent surface structures. The plant's risk of infection is reduced by removing water from the surface. Additionally, removing dust from leaf surfaces reduces the risk of salt damage and protects the plant from overheating [7, 11].

The concept of self-cleaning was initially inspired by the superhydrophobic property observed in certain plant leaves. Among them, the lotus leaf is the most famous as it allows water droplets to effortlessly roll off its surface, keeping it clean [3–5]. Another illustration of superhydrophobicity in nature can be found in cabbage leaves and Indian cress plants [9].

The spatial arrangement of topographical figures on the plant surfaces allows water-drops and surface contaminants to roll off effortlessly. Water drops can slide off grass leaves transiently, thanks to lines of topographical protuberances that run along the main leaf direction. Similar characteristics can be seen in rice, taro, and Indian canna leaves. Rice leaves exhibit an intriguing property accepted as anisotropic (diverse properties in different directions) wettability. The surface of the leaves is composed of a binary micro/nanostructure, and the papillae, which have an average diameter of 5–8 millimeters, are placed in a one-dimensional sequence parallel to the leaf edge (**Figure 2a–d**). Because of the peculiar anisotropic arrangement of the papillae, water droplets can roll along rice leaves easier in the way of the edge than in a perpendicular direction [2, 7, 13]. Quite a few research teams have tried to replicate these conditions by designing surfaces with uniformly anisotropic wettability models. Yoshimitsu and colleagues, for instance, provided proof that water flows more easily on substrates with pillar lines parallel to them than it does when moving orthogonally (5, 10).

While one dominant length scale is primarily responsible for the physical roughness effects that give rice leaves their ability to self-clean, there are other circumstances where multiple length scales are present and work together. The lotus leaf is a prime example of this, as it exhibits various roughness length scales on its surface, ranging from nanometers to microscopic ones, instead of just one. On the surface of the lotus leaf, water droplets remove contaminated particulate compared to other hydrophobic plant surfaces (**Figure 3**). The presence of assorted length sizes of roughness reduces the tendency of dirt particulate matter's adhesion on a leaf. The low and high magnification scanning electron microscope (SEM) of its surface is demonstrated in **Figure 3b** and **c**. Thus, facilitating the water droplets passing by to carry them away. Gao *et al*. offer a detailed explanation of how multiple-level roughness affects wettability in this scenario [7, 15, 16].

Figure 2.
a) Rice leaves, which exhibit extraordinary water repellence on their upper and downsides. (b) A few water droplets float on a rice leaf. (c, d) The scanning electron micrograph (SEM) of the rice leaf surface image of the upper leaf side. Source: [12].

Figure 3.
(a) Lotus leaves, which exhibit extraordinary water repellence on their upper side. (b) Scanning electron microscopy (SEM) image of the upper leaf side prepared by 'glycerol substitution' shows the hierarchical surface structure consisting of papillae, wax clusters, and wax tubules. (c) Wax tubules on the upper leaf side. (d) Upper leaf side after critical-point (CP) drying. The wax tubules are dissolved; thus, the stomata are more visible. Tilt angle 15°. (e) The leaf underside (CP dried) shows convex cells without stomata. Source: [14].

Young and his colleagues have put forward wetting models that shed light on how lotus leaves clean themselves. Young's equation for the WCA is shown below (1), where θ_0 is the WCA of the droplet on the surfaces, γ_{SA} and γ_{SL} display the surface energies of the solid against air and liquid, respectively, and γ_{LA} is the liquid-air interfacial energy (**Figure 4**).

$$cos\,\theta_0 = \frac{\gamma_{SA} - \gamma_{SL}}{\gamma_{LA}} \quad (1)$$

Using Young's equation, a water droplet on a flat, homogeneous surface can successfully predict the WCA. Wenzel's Eq. (2) can be employed to determine the WCA if the surface is rough, and the actual surface area exceeds the flat expected area (**Figure 5**), where Rf is the ratio of the actual surface area to its flat expected area.

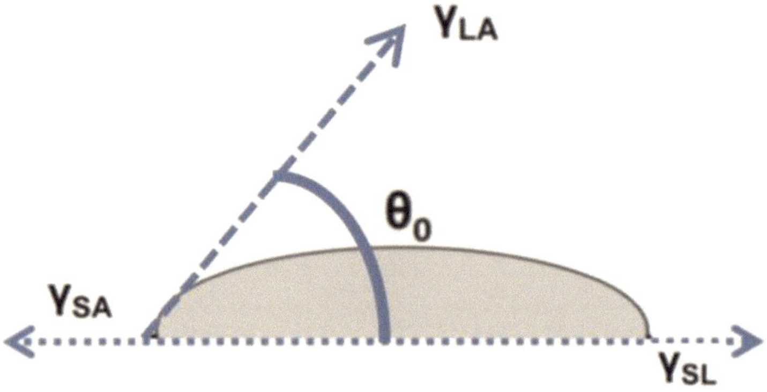

Figure 4.
Liquid drop schematic to show the values in Young equation.

$$\cos\theta_w = R_f \cos\theta_0 \tag{2}$$

For a rough surface, $R_f > 1$, which means that for rough surfaces, hydrophilic surfaces grow more hydrophilic, and hydrophobic surfaces get more hydrophobic. The Cassie Eq. (3) can also be derived for heterogeneous surfaces made of two fractions.

$$\cos\theta = f_1 \cos\theta_1 + f_2 \cos\theta_2 \tag{3}$$

where f_1 and f_2 are the fractional area with CA θ_1 and θ_2, respectively.

The Cassie-Baxter Eq. (4) can be applied to a composite interface that consists of the solid-liquid fraction ($f1 = f_{SL}$ and $\theta_1 = \theta_0$) and the liquid–air fraction ($f_2 = f_{LA}$ and $\cos\theta_{21} = -1$) (**Figure 5**).

$$\cos\theta_{CB} = R_f \cos\theta_0 + f_{LA}\left(R_f \cos\theta_0 + 1\right) \tag{4}$$

Based on these suggested models, a surface can develop the potential to self-clean itself by manipulating surface microstructures to encourage the free, spontaneous

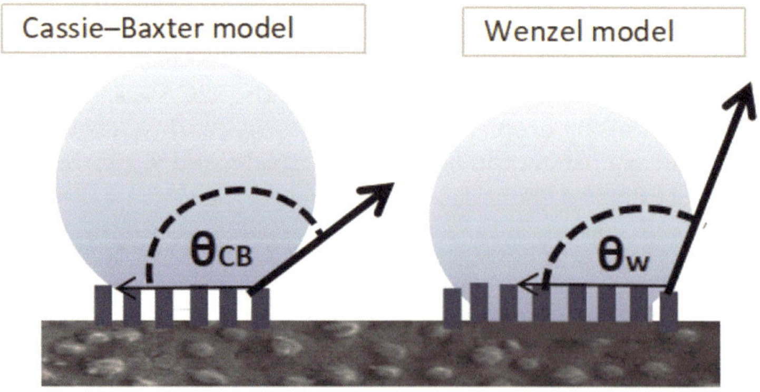

Figure 5.
Droplets display on rough surfaces; Cassie–Baxter model and Wenzel model (adapted from [6, 17, 18]).

movement of liquid droplets on the surface, which enables the removal of contaminants from the surfaces. Therefore, ensuring that droplets can flow or roll off smoothly from the connected surface with no resistance will be the main objective for a surface to accomplish self-cleaning. To enable free liquid droplet movement, natural or bio-surfaces primarily increase or decrease the droplets' contact angles. To create this unrestricted movement, the surface energy of the three phases—liquid, air, and solid—is controlled [3].

Droplets move freely across the surface of the leaf when the WCA is close to 180° (e.g., lotus leaves), dissolving dirt components along the droplets' motion in the liquid [17].

The opposite strategy (super-hydrophilic surfaces) is used by a number of other natural systems, for instance, pitcher plants, *Anubias barteri,* and *Heliamphora nutans*. Water droplets spread quickly on a super-hydrophilic surface; a water film is formed on them. Additionally, the tiny WCA droplets can be used as a sharp knife to remove contaminants from the surface and separate dirt. Therefore, super-hydrophilic surfaces have also attracted interest for their self-cleaning surface design. For example, the pitcher plant's slick surface is thought to be one of its most crucial characteristics for catching insects. Due to the surface microtopography and hygroscopic nectar secretion, the peristome exhibits superhydrophilicity. As a result, when the air is humid, stable water films form. The oils on the insects' feet make this water film repellent, causing insects that tread on it to slip over the rim and into the pitcher plant's bottom, which is filled with digestive liquid [2, 3, 6].

4. Reassembling and formulation

Recently, Lotus leaves have inspired many synthetic superhydrophobic surfaces fabricated by applying self-cleaning coatings on different substrata such as textiles, tiles, and glass [19].

Super-hydrophobicity can be produced in one of two ways: either by chemically altering a hierarchically structured surface with low surface energy material or; and by building micro/ nanostructures on hydrophobic substrates. Numerous techniques for synthesizing these surfaces have been documented in published literature, including immersion, electrochemical deposition, emulsion, electrospinning, phase separation, Chemical vapor deposition (CVD), spray coating, wet chemical reaction, 3D-printing, etc. [20, 21].

Zhao and colleagues [22] used poly(dimethylsiloxane) (PDMS) replication material and a natural rice leaf as a template to fabricate a rice leaf replica on an Au surface. Gao *et al.* [23] used a three-step process (I) negative replication of a rice leaf using PDMS; (II) positive replication of a PDMS template using poly (N-isopropylacrylamide) (PNIPAAm) in hot water; and (III) separation of the replicated PNIPAAm film from the template. The synthesized rice leaf replica showed good, responsive anisotropic wettability as a result of the polymer structure's change in thermal responsiveness. The WCA was 119 ± 10° at 50°C and dramatically decreased to 77 ± 9° at 20°C when measured in the parallel direction of the grooves. However, the WCA was 87 ± 9° at 50°C and 49 ± 6° at 20°C in the perpendicular direction. Smart surfaces are, therefore, those that respond to stimuli, such as surfaces that are wettable and switch between pH, temperature, light, and electric field. Combining the responsiveness of these surfaces with self-cleaning surfaces is a strategy for new applications [2].

In the past, low surface energy materials were used to create superhydrophobic surfaces. One of the most popular fluorine-containing polymers with superhydrophobicity is polytetrafluoroethylene (PTFE). By adding organic groups like CF_3, CH_3, and CH_2, etc., the hydrophobicity of materials can be increased. Superhydrophobic materials can also be produced using Polyesters and Polyurethanes (PUs), Polyethylene, Polystyrene, Polyvinyl chloride, and Polydimethylsiloxane (PDMS). However, fluorinated hydrocarbon compounds, which pose serious risks to both human health and the environment, are frequently used in the development of these surfaces [19, 24].

Using hydrophobic dual-scaled SNPs and PDMS via a hybrid route combining soft imprinting and spin-coating, Ghasemlou and colleagues [19] discovered a straightforward and environmentally friendly method to fabricate multifunctional superhydrophobic surfaces on SPC films with a well-defined micro - nanoscale hierarchical structure. Direct soft-imprinting lithography on starch -polyhydroxyurethane - cellulose nanocrystal (SPC) films resulted in micro-scaled features that resembled lotus leaf pillar architecture. A thin layer of low-surface energy material, poly (dimethylsiloxane) (PDMS), was assembled over these microstructures using a spin-coating technique. Functional silica nanoparticles (V-SNPs) were created by grafting silica nanoparticles (SNPs) with vinyltriethoxysilane (VTES). These V-SNPs were then used to create superhydrophobic coatings. A further modification of the DPMS@SPC film with V-SNPs allowed the interlocking of V-SNP microparticles within the cross-linked PDMS network. It was claimed that an extremely hydrophobic surface was synthesized, which had a water contact angle (WCA) of 150° and a sliding angle (SA) of 10°. He pointed out that when compared to uncoated films, the water vapor transmission rate of the PDMS/V-SNP@SPC films decreased by 52%. These findings suggested that the coating effectively protected the film substrate from moisture and gave it good hydrophobicity. The film surface with coating displayed exceptional mechanical strength as they could withstand severe knife scratches, rubbing with fingers, jet-water impact, 20 cycles of sandpaper abrasion tests, and ten repetitions of tape-peeled tests without losing their superhydrophobicity. When artificial dust and various food liquids were removed from the surfaces, self-cleaning behavior was also observed [19].

Bohn [25] and Wong *et al.* [26] were motivated by the system observed on the slippery surface of the pitcher plant. They created "slick liquid-infused porous surface (s)" (SLIPS), which they referred to as synthetic liquid-repellent surfaces. A micro/nano-porous substrate holds the film of a lubricating liquid in place in SLIPS. The materials for SLIPS were selected according to the following criteria: (I) the lubricating liquid must moisten the substrate, wick into it, and securely attach to it; (II) the solid must be preferentially moistened by the liquid one wants to repel; and (III) the immiscible nature of the moistening and impinging test liquids [2].

5. Application in food, pharmaceutical, and related industries

In recent years, bio-inspired materials and surfaces with self-cleaning properties have been developed rapidly due to the advancement in nanotechnology and engineering nanoscience [27]. In addition to lotus leaves, it is possible to create materials with superhydrophobic and superhydrophilic properties by imitating the structures of plants like rice leaves and pitcher plants. [2]. Different fabrication techniques have been utilized to create biomimetic self-cleaning coatings by modifying rough surfaces with low-surface energy materials or roughening the surface of low-surface energy materials [28]. Several self-cleaning products have been commercialized based on these techniques, such as construction materials, glasses, solar panels, and windows. Besides, self-cleaning materials and surfaces are drawing attention for their potential application in the food, pharmaceutical, and detergent industries. A summary of recently reported applications is provided in **Table 2**.

In the food industry, research has been focused on applying self-cleaning coatings on food packaging materials and food processing equipment. For instance, Cai *et al.* [29] fabricated a food packaging material - hydrophobic starch nanofibrous film (SNF) by assembling stearic acid (STA) onto SNF to create a hierarchical micro-nano structure inspired by lotus leaves. This hydrophobic SNF exhibits self-cleaning properties by enabling water to roll freely in all directions. The self-assembled coated SNF from STA shows great potential as a food packaging material due to its biodegradable, edible, and waterproof properties. Additionally, its self-cleaning capabilities aid in preventing fouling on the packaging surfaces. In addition, self-cleaning surfaces are demonstrated to exhibit antibacterial and antimicrobial properties by reducing the adhesion of bacteria and microorganisms to the surface. For instance, the nanocomposite surface coating developed by Yoon *et al.* [31] demonstrated effective inhibition of bacterial adhesion with 80% fewer bacteria adhered on the metal surface in comparison to those on the uncoated surface. The superhydrophobic surface coating was fabricated using a cluster of carbon nanotubes (CNTs) with low surface energy nanoscale roughness and polytetrafluoroethylene (PTFE). This CNT-PTFE composite was then spray-coated on stainless-steel plates to exhibit self-cleaning properties via the lotus effect.

The application of self-cleaning nanocomposite coating on food contact surfaces (e.g., food processing equipment) is an ideal solution to reduce bacterial adhesion and biofilm formation, leading to a lower risk of cross-contamination and food safety hazards. Besides, the nanocomposite coatings also help reduce the use of water and chemicals for the cleaning process.

It is also feasible to employ such self-cleaning materials or surfaces within the pharmaceutical industry to improve the safety and efficiency of drug manufacturing processes. For instance, filtration is the process widely used in industry for the concentration, separation, and purification of chemical solutions. However, filtration with conventional filters may affect the quality and safety of pharmaceutical products, such as the potential adsorption of drug components to the filter membranes, which reduces the concentration of adsorbed components and the leaching of filter biofilms into the pharmaceutical products, which may lead to undesirable toxicity [42, 43]. Therefore, filters with self-cleaning coatings shall be adopted in the filtration process to prevent the adsorption of drug components, reduce bacterial adhesion, and prevent biofilm formation.

Other than filters, self-cleaning materials can also be applied to the surface coatings of production equipment. Hence, the equipment can be operated continuously without the need for manual cleaning and sterilization. Such a continuous production system can help to increase the efficacy of production and reduce the risk of cross-contamination.

Self-cleaning materials also play a crucial role in the detergent industry to make cleaning easier and more convenient for consumers and mitigate environmental pollution problems. Traditional detergents containing petroleum-based surfactants are effective at removing dirt and stains, but they are also hazardous to human health and the environment [44]. Therefore, there has been a trend towards developing detergents that are more sustainable and eco-friendlier. For instance, Yang

Area	Material	Method	Self-cleaning action	Application	Reference
Food Packaging/ Food Processing Equipment	Starch nanofibrous film (SNF)/ stearic acid (STA)	Temperature-assisted electro spinning	Hydrophobic	Biodegradable, self-cleaning food packaging.	[29]
	Multi-walled Carbon nanotubes/ perfluorooctanesulfonyl fluoride	Surface fluorination	Superhydrophobic with antibacterial	Antibacterial, self-cleaning food packaging.	[30]
	Carbon nanotubes (CNTs)/ Polytetrafluoroethylene (PTFE)	Spray coating	Superhydrophobic with antibacterial	Reduce bacterial adhesion on food processing and packaging equipment.	[31]
	Lignin-coated cellulose nanocrystal (L-CNC)/ polyvinyl alcohol	Spray coating and chemical vapor deposition	Superhydrophobic	Self-cleaning food packaging with excellent abrasion resistance and nontoxicity.	[32]
	SiO_2 nanoparticles/ polydimethylsiloxane/ varnish	Modification of the overprint varnish	Superhydrophobic with anti-frosting	Waterproof, easy cleaning, anti-frost paper food packaging.	[33]
	STA modified organic montmorillonite/ poly(dimethylsiloxane)	Self-assembly	Superhydrophobic with anti fouling	Self-cleaning surface to reduce liquid-food residue	[34]
Fabrics/Textiles	Titania nanosols	Sol-gel	Photocatalytic	Self-cleaning cotton with easy stain removal	[35]
	Poly(methylmethacrylate) nanofibers/ ZnO nanorods/Ag nanoparticles	Electrospinning	Photocatalyti	Self-cleaning protective clothing with antibacterial and antiviral properties	[36]
Glass	3-Aminopropytriethoxysilane (APTS)-modified hollow silica nanoparticle	Dip coating and chemical vapor deposition	Superhydrophobic	Self-cleaning glass with high transparency	[37]
	TiO_2 nanoparticles	Sol-gel	Superhydrophilicity	Self-cleaning glass with antifogging	[38]
	TiO_2/SiO_2 nanoparticle	Layer by layer deposition	Superhydrophilicity	Self-cleaning glass with antifogging and anti-reflection	[39]
Wastewater treatment	SiO_2 nanoparticles	Dip coating	Superhydrophobic and superoleophilic	Self-cleaning sponge for oil-water separation	[40]
Solar cells	SiO_2 nanoparticles array/poly(ethylene terephthalate)	Dry coating and chemical vapor deposition	Hydrophobic and oleophobic	Self-cleaning solar cells with light scattering properties	[41]

Table 2.
Applications of self-cleaning materials in various industries.

et al. [45] have developed an eco-friendly, non-toxic detergent using the halloysite clay nanotube (HNT), which demonstrated high cleaning capacity to remove different strains from various substrates. These hydrophilic nanoparticles can be potentially designed to exhibit superhydrophilicity so the dirt can be removed easily with the quick spreading of water, leading to improved decontamination and cleaning effects [46].

Besides, the self-cleaning nanoparticles can be designed to attach themselves to the surfaces and prevent future dirt accumulation, resulting in extended cleanliness of surfaces over long periods of time.

These nanoparticles have been well-developed to produce self-cleaning materials with anti-fouling and antimicrobial properties, but they are also associated with some drawbacks in toxicity, stability, and durability. An example of a concern related to nanoparticles is the potential for them to dissociate from surface coatings and be inhaled by humans, which could have unintended consequences. New fabrication technologies, such as encapsulation, should be taken into consideration to overcome those shortcomings.

Brown and Bhushan [47] fabricated the nanoparticles-encapsulated surfaces using SiO2 nanoparticles. The nanoparticles-encapsulated surface exhibited good self-cleaning properties because it removed more than 90% of the contaminants on the surface. Besides, it demonstrated high durability with no noticeable wear scar on the treated surfaces after 200 cycles of tribometer wear experiment [47]. The highly durable surface can be attributed to the high hardness of SiO2 nanoparticles for wear resistance and the encapsulation technique.

The encapsulation technique directly incorporated nanoparticles into the surface during the softening process of polymer material, compared to other techniques in which nanoparticles are employed as a coating. The durability of the self-cleaning surface typically relies on the adhesion of nanoparticles to the coating materials [48]. Therefore, the nanoparticles-encapsulation technique will be one of the most anticipated alternatives for fabricating the self-cleaning surface due to the improved mechanical durability and stability.

In summary, self-cleaning technology has the potential to improve performance and efficiency in the food, pharmaceutical, and detergent industries. Further research is needed to explore new fabrication techniques, such as nanoparticle encapsulation, in order to create materials that are self-cleaning and have excellent durability and stability.

6. Design a self-cleaning structure

As described in this chapter, understanding the basic principles of self-nano-cleaning surface structures can be used as a guide and applied to various applications. One possible solution for creating a self-cleaning surface with various nano/micro surfaces is to utilize advanced 3D printing technology. It enables the fabrication of biomimetic functional surfaces by digitally controlling the complex hierarchical microstructures. An interesting example of biomimicry is the fabrication of artificial hairs resembling the eggbeater heads of *Salvinia molesta* leaves, which were produced by the immersed surface accumulation 3D printing process [10]. This involves designing the microstructures of bio-inspired models using computer-aided design software, which was subsequently sliced into layers and mapped onto the object surfaces. The photopolymers were then cured layer by layer through the projection of UV light onto the surface. Finally, a functional surface with designed microstructures is fabricated. The results show that the fabricated functional surfaces have controllable hydrophobic properties. In general, 3D printing technologies open a new door for the design and assembly of functional textures with desired properties. It shows a promising future around the fabrication of functional surfaces area with hydrophobic and hydrophilic properties. However, it is important to acknowledge that this technology still has certain limitations. One potential drawback is that this method tends to be slower than some other fabrication methods. Additionally, it's important to note that only certain types of photopolymers are suitable for use with this method. Unfortunately, these materials may not be food-safe or biocompatible, which can limit their applications.

7. Conclusions

The self-cleaning surface found in plant structures is ultimately determined by two factors: 1) the diverse textures of the leaf's surface and 2) the presence of various chemicals, waxes, and other components that affect its surface energy level. The surface can vary from highly rough to highly smooth, and the coverage of wax on leaves can range from high-surface energy to low-surface energy. This can result in surfaces that are super-hydrophobic or super-hydrophilic. This phenomenon is the basic knowledge to understand the mechanism and reassembling of the self-nano/micro-cleaning surface [18]. To design and make the self-cleaning surface with different nano/micro surfaces, 3D printing can be one of the accessible solutions. The advantage of 3D printing machinery is that the surface can be designed and assembled according to the required texture with various degrees of roughness/smoothness (low to high).

Acknowledgements

The authors would like to thank Dr. Ranjeet Singh for editing and adding valuable feedback.

The authors would also like to thank the Faculty of Science Equity Diversity & Inclusion Committee, Monash University, for approving the "Advance Women's Success Grant" for publishing this book chapter.

Author details

Mojdeh Nakhaei, Jing Ying Chong, Yunlong Tang and Shahnaz Mansouri*
Monash University, Melbourne, Australia

*Address all correspondence to: shahnaz.mansouri@monash.edu

References

[1] Barthlott W, Neinhuis C. Purity of the sacred lotus, or escape from contamination in biological surfaces. Planta. 1997;**202**:1-8. DOI: 10.1007/s004250050096

[2] Nishimoto S, Bhushan B. Bioinspired self-cleaning surfaces with superhydrophobicity, superoleophobicity, and superhydrophilicity. RSC Advances. 2013;**3**(3):671-690. DOI: 10.1039/C2RA21260A

[3] Xu Q, Zhang W, Dong C, Sreeprasad TS, Xia Z. Biomimetic self-cleaning surfaces: Synthesis, mechanism and applications. Journal of the Royal Society Interface. 2016;**13**(122): 20160300. DOI: 10.1098/rsif.2016.0300

[4] Koch K, Bhushan B, Barthlott W. Diversity of structure, morphology and wetting of plant surfaces. Soft Matter. 2008;**4**(10):1943-1963. DOI: 10.1039/B804854A

[5] Watson GS, Gellender M, Watson JA. Self-propulsion of dew drops on lotus leaves: A potential mechanism for self cleaning. Biofouling. 2014;**30**(4):427-434. DOI: 10.1080/08927014.2014.880885

[6] Zhang M, Feng S, Wang L, Zheng Y. Lotus effect in wetting and self-cleaning. Biotribology. 2016;**5**:31-43. DOI: 10.1016/j.biotri.2015.08.002

[7] Genzer J, Marmur A. Biological and synthetic self-cleaning surfaces. MRS Bulletin. 2008;**33**(8):742-746. DOI: 10.1557/mrs2008.159

[8] Yoshimitsu Z, Nakajima A, Watanabe T, Hashimoto K. Effects of surface structure on the hydrophobicity and sliding behavior of water droplets. Langmuir. 2002;**18**(15):5818-5822. DOI: 10.1021/la020088p

[9] Prianka TR, Subhan N, Reza HM, Hosain MK, Rahman MA, Lee H. Recent exploration of bio-mimetic nanomaterial for potential biomedical applications. Materials Science and Engineering: C. 2018;**93**:1104-1115. DOI: 10.1016/j.msec.2018.09.012

[10] Yang Y, Li X, Zheng X, Chen Z, Zhou Q, Chen Y. 3D-printed biomimetic super-hydrophobic structure for microdroplet manipulation and oil/water separation. Advanced Materials. 2018; **30**(9):1704912. DOI: 10.1002/adma.201704912

[11] Neinhuis C, Barthlott W. Characterization and distribution of water-repellent, self-cleaning plant surfaces. Annals of Botany. 1997;**79**(6): 667-677. DOI: 10.1006/anbo.1997.0400

[12] Zhou ZL, Cao C, Cao LD, Zheng L, Xu J, Li FM et al. Evaporation kinetics of surfactant solution droplets on rice (Oryza sativa) leaves. PLoS ONE. 2017; **12**(5): e0176870. https://doi.org/10.1371/journal.pone.0176870

[13] Guo Z, Liu W. Biomimic from the superhydrophobic plant leaves in nature: Binary structure and unitary structure. Author Links Open Overlay Panel. 2007; **172**(6):1103-1112. DOI: 10.1016/j.plantsci.2007.03.005

[14] Ensikat HJ, Ditsche-Kuru P, Neinhuis C, Barthlott W. Superhydrophobicity in perfection: the outstanding properties of the lotus leaf [Internet]. Beilstein Journal of Nanotechnology. Beilstein Institut; 2011; 2:152-161. Available from: http://dx.doi.org/10.3762/bjnano.2.19

[15] Li X, Gonga F, Liua D, Hea S, Yuana H, Daia L, et al. A lotus leaf based random laser. Organic Electronics. 2019;

69:216-219. DOI: 1016/j.orgel.2019.
03.032

[16] Genzer J, Efimenko K. Recent developments in superhydrophobic surfaces and their relevance to marine fouling: A review. Biofouling. 2006; **22**(5):339-360. DOI: 10.1080/ 08927010600980223

[17] Khojasteh D, Kazerooni M, Salarian S, Kamali R. Droplet impact on superhydrophobic surfaces: A review of recent developments. Journal of Industrial and Engineering Chemistry. 2016;**42**:1-14. DOI: 10.1016/j. jiec.2016.07.027

[18] Kota AK, Kwon G, Tuteja A. The design and applications of superomniphobic surfaces. NPG Asia Materials. 2014;**6**:e109. DOI: 10.1038/ am.2014.34

[19] Ghasemlou M, Le PH, Daver F, Murdoch BJ, Ivanova EP, Adhikari B. Robust and eco-friendly superhydrophobic starch nanohybrid materials with engineered lotus leaf mimetic multiscale hierarchical structures. ACS Applied Materials & Interfaces. 2021;**13**(30):36558-36573. DOI: 10.1021/acsami.1c09959

[20] Kumar A, Nanda D. Methods and fabrication techniques of superhydrophobic surfaces. In: Samal S, Mohanty S, Nayak SK, editors. Superhydrophobic Polymer Coatings: Fundamentals, Design, Fabrication, and Applications. 1st edition. US: Elsevier; 2019. pp. 43-75. DOI: 10.1016/B978-0-12-816671-0.00004-7

[21] Atwah AA, Almutairi MD, He F, Khan MA. Influence of printing parameters on self-cleaning properties of 3D printed polymeric fabrics. Polymers. 2022;**14**(15):3128. DOI: 10.3390/ polym14153128

[22] Zhao W, Wang L, Xue Q. Fabrication of low and high adhesion hydrophobic Au surfaces with micro/nano-biomimetic structures. The Journal of Physical Chemistry C. 2010;**114**(26):11509-11514. DOI: 10.1021/jp102052e

[23] Gao J, Liu Y, Xu H, Wang Z, Zhang X. Biostructure-like surfaces with thermally responsive wettability prepared by temperature-induced phase separation micromolding. Langmuir. 2010;**26**(12):9673-9676. DOI: 10.1021/ la100256b

[24] Shirsat V, Abhinayaa N, Sabnis A. Superhydrophobic polymer coating: A design perspective. In: Samal S, Mohanty S, Nayak SK, editors. Superhydrophobic Polymer Coatings: Fundamentals, Design, Fabrication, and Applications. 1st edition. US: Elsevier; 2019. pp. 31-42. DOI: 10.1016/B978-0-12-816671-0.00003-5

[25] Bohn H. From the cover: Insect aquaplaning: Nepenthes pitcher plants capture prey with the peristome, a fully wettable water-lubricated anisotropic surface. Proceedings of the National Academy of Sciences. 2004;**101**(39): 14138-14143. DOI: 10.1073/pnas. 040588510

[26] Wong T-S, Kang SH, Tang SK, Smythe EJ, Hatton BD, Grinthal A. Bioinspired self-repairing slippery surfaces with pressure-stable omniphobicity. Nature. 2011;**477**(7365): 443-447. DOI: 10.1038/nature10447

[27] Zhang J, Wang J, Zhao Y, Xu L, Gao X, Zheng Y. How does the leaf margin make the lotus surface dry as the lotus leaf floats on water? Soft Matter. 2008;**4**(11):2232-2237. DOI: 10.1039/ B807857B

[28] Ganesh VA, Raut HK, Nair AS, Ramakrishna S. A review on self-

cleaning coatings. Journal of Materials Chemistry. 2011;**21**(41):16304-16322. DOI: 10.1039/C1JM12523K

[29] Cai J, Zhang D, Zhou R, Zhu R, Fei P, Zhu ZZ, et al. Hydrophobic interface starch nanofibrous film for food packaging: From bioinspired design to self-cleaning action. Journal of Agricultural and Food Chemistry. 2021; **69**(17):5067-5075. DOI: 10.1021/acs.jafc.1c00230

[30] Song K, Gao A, Cheng X, Xie K. Preparation of the superhydrophobic nano-hybrid membrane containing carbon nanotube based on chitosan and its antibacterial activity. Carbohydrate Polymers. 2015;**130**: 381-387. DOI: 10.1016/j.carbpol.2015.05.023

[31] Yoon SH, Rungraeng N, Song W, Jun S. Superhydrophobic and superhydrophilic nanocomposite coatings for preventing Escherichia coli K-12 adhesion on food contact surface. Journal of Food Engineering. 2014;**131**: 135-141. DOI: 10.1016/j.jfoodeng.2014.01.031

[32] Huang J, Lyu S, Fu F, Wu Y, Wang S. Green preparation of a cellulose nanocrystals/polyvinyl alcohol composite superhydrophobic coating. RSC Advances. 2017;**7**(33):20152-20159. DOI: 10.1039/C6RA27663F

[33] Chen W, Wang X, Tao Q, Wang J, Zheng Z, Wang X. Lotus-like paper/paperboard packaging prepared with nano-modified overprint varnish. Applied Surface Science. 2013;**266**:319-325. DOI: 10.1016/j.apsusc.2012.12.018

[34] Sun Y, Liu R, Xu J, Sun Y, Gong J, Long L. A durable and environmental friendly superhydrophobic coatings with self-cleaning, anti-fouling performance for liquid-food residue reduction. Polymer Engineering & Science. 2023; **63**(4):1274-1288. DOI: 10.1002/pen.26282

[35] Abidi N, Cabrales L, Hequet E. Functionalization of a cotton fabric surface with titania nanosols: Applications for self-cleaning and UV-protection properties. ACS Applied Materials & Interfaces. 2009;**1**(10): 2141-2146. DOI: 10.1021/am900315t

[36] Karagoz S, Kiremitler NB, Sarp G, Pekdemir S, Salem S, Goksu AG, et al. Antibacterial, antiviral, and self-cleaning mats with sensing capabilities based on electrospun nanofibers decorated with ZnO nanorods and Ag nanoparticles for protective clothing applications. ACS Applied Materials & Interfaces. 2021; **13**(4):5678-5690. DOI: 10.1021/acsami.0c15606

[37] Xu L, He J. Fabrication of highly transparent superhydrophobic coatings from hollow silica nanoparticles. Langmuir. 2012;**28**(19):7512-7518. DOI: 10.1021/la301420p

[38] Zorba V, Chen X, Mao SS. Superhydrophilic TiO 2 surface without photocatalytic activation. Applied Physics Letters. 2010;**96**(9):093702. DOI: 10.1063/1.3291667

[39] Lee D, Rubner MF, Cohen RE. All-nanoparticle thin-film coatings. Nano Letters. 2006;**6**(10):2305-2312. DOI: 10.1021/nl061776m

[40] Latthe SS, Sutar RS, Kodag VS, Bhosale AK, Kumar AM, Sadasivuni KK, et al. Self–cleaning superhydrophobic coatings: Potential industrial applications. Progress in Organic Coatings. 2019;**128**:52-58. DOI: 10.1016/j.porgcoat.2018.12.008

[41] Jeong E, Zhao G, Song M, Yu SM, Rha J, Shin J, et al. Simultaneous improvements in self-cleaning and

light-trapping abilities of polymer substrates for flexible organic solar cells. Journal of Materials Chemistry A. 2018; **6**(5):2379-2387. DOI: 10.1039/ C7TA09351A

[42] Bin T, Kulshreshtha AK, Al-Shakhshir R, Hem SL. Adsorption of benzalkonium chloride by filter membranes: Mechanisms and effect of formulation and processing parameters. Pharmaceutical Development and Technology. 1999;**4**(2):151-165. DOI: 10.1081/pdt-100101350

[43] Pillai SA, Chobisa D, Urimi D, Ravindra N. Filters and filtration: A review of mechanisms that impact cost, product quality and patient safety. Journal of Pharmaceutical Sciences and Research. 2016;**8**(5):271-278. https:// www.proquest.com/docview/ 1792593649?pq-origsite=gscholar& fromopenview=true

[44] Do DN, Dang TT, Le QT, Lam TD, Bach LG, Nguyen DC, et al. Extraction of saponin from gleditsia peel and applications on natural dishwashing liquid detergent. Materials Today: Proceedings. 2019;**18**:5219-5230. DOI: 10.1016/j.matpr.2019.07.522

[45] Yang X, Cai J, Chen L, Cao X, Liu H, Liu M. Green detergent made of halloysite nanotubes. Chemical Engineering Journal. 2021;**425**: 130623. DOI: 10.1016/j.cej.2021. 130623

[46] Zhang L, Zhao N, Xu J. Fabrication and application of superhydrophilic surfaces: A review. Journal of Adhesion Science and Technology. 2014;**28**(8–9): 769-790. DOI: 10.1080/01694243. 2012.697714

[47] Brown PS, Bhushan B. Durable superoleophobic polypropylene surfaces. Philosophical Transactions of the Royal Society A: Mathematical, Physical and Engineering Sciences. 2016; **374**(2073):20160193. DOI: 10.1098/ rsta.2016.0193

[48] Martin S, Brown PS, Bhushan B. Fabrication techniques for bioinspired, mechanically-durable, superliquiphobic surfaces for water, oil, and surfactant repellency. Advances in Colloid and Interface Science. 2017;**241**:1-23. DOI: 10.1016/j.cis.2017.01.004

Chapter 3

Self-Cleaning Surfaces of Polyurethanes

Kamlesh Panwar, Neelam Vyas, Roopali Rai and Shreedhar Bhat

Abstract

In this urbanized world, people have limited time and access to labors to clean the items one is associated with. Self-cleaning of the items which humans use every day or occasional is more sustainable for long term and is also one of the most important functionalities for improved esthetics, performance, hygiene, and satisfaction. Various approaches have been widely explored to impart self-cleaning properties to different substrates using different chemistries of surface modifications. The current chapter gives an overview of the various mechanisms for self-cleaning including super-hydrophobicity, super-hydrophilicity and photocatalysis with more emphasis on polyurethane origin. Polyurethanes have been widely explored for self-cleaning properties by introducing super-hydrophobicity via incorporation of nano-roughness or low energy functionalities or by introducing photocatalytic property by incorporating photocatalytic nanoparticles. The chapter also provides a connect to the applications of such polyurethane surfaces. Thus, these self-cleaning polyurethanes may find applications in the fields of anti-fogging, anti-icing, anti-reflection, corrosion resistance, drag reduction, sensors, solar cells, and textiles.

Keywords: Photocatalysis, polyurethanes, super-hydrophilic, super-hydrophobic, applications

1. Introduction

The 20th century saw a revolution in plastic and polymer innovation and production. Scientists in both academic and industrial laboratories were synthesizing new monomers from affordable and abundant raw materials. The terms 'versatility' and 'popularity' apply to a limited number of polymers available in the world, polyurethanes are one among these limits. Polyurethanes are a versatile class of polymers with great control over their physicochemical properties based on the chemical composition. From insulation to surf boards, from car airbags to window sealants, polyurethanes are everywhere in our daily lives.

Polyurethane (PU) was first introduced by a German professor, Dr. Otto Bayer and his co-workers in the 1940s [1], and has been applied in a very broad range of commercial and industrial fields due to its unique combination of unusual features including excellent mechanical strength, good abrasion resistance, toughness, low temperature flexibility, corrosion resistance, processability, etc. The basic repetitive unit in PUs is the urethane group (—NHCOO—), which is produced from the reaction between isocyanate (—NCO), polyols (—OH), and other additives [2]. Segmented polyurethanes are composed of two blocks: the soft segment is formed by a macrodiol (polyether or polyester diol), and the hard segment is composed by a diisocyanate and a low molecular weight chain extender or crosslinkers [3]. Their growing success and increased use are further boosted by the fact that they are affordable, safe and recyclable, and these qualities make them the product of choice for manufacturers and retailers all over the world.

Self-cleaning surfaces are a class of materials with the inherent ability to remove any debris or bacteria from their surfaces in a variety of ways. The self-cleaning functionality of these surfaces are commonly inspired by natural phenomena observed in lotus leaves, gecko feet, and water striders to name a few. The first instance of a self-cleaning surface was created in 1995 [4]. Paz et al. [4] created a transparent titanium dioxide (TiO_2) film that was used to coat glass and provide the ability for the glass to self-clean. The first commercial application of this self-cleaning surface, Pilkington Activ, was developed by Pilkington glass in 2001. This chapter deals with the fundamentals of bio-inspired self-cleaning phenomenon followed by different approaches adopted to achieve such surfaces. Furthermore, recent advancement in polyurethane self-cleaning surfaces and applications have also been analyzed and discussed.

2. Fundamental theories relevant to self-cleaning

2.1 Typical wetting models

The ability of a surface to self-clean commonly depends on the hydrophobicity or hydrophilicity of the surface. Whether cleaning aqueous or organic matter from a surface, water plays an important role in the self-cleaning process. The mechanism of self-cleaning is majorly explained in terms of contact angle of water droplets on the coating surface. Based on the water contact angle, the substrates are categorized into four types- super-hydrophilic, hydrophilic, hydrophobic, and super-hydrophobic, as shown in **Figure 1**. The surface is hydrophilic when the contact angle of water droplets on the surface is below 90° and super-hydrophilic in case the water contact angle is

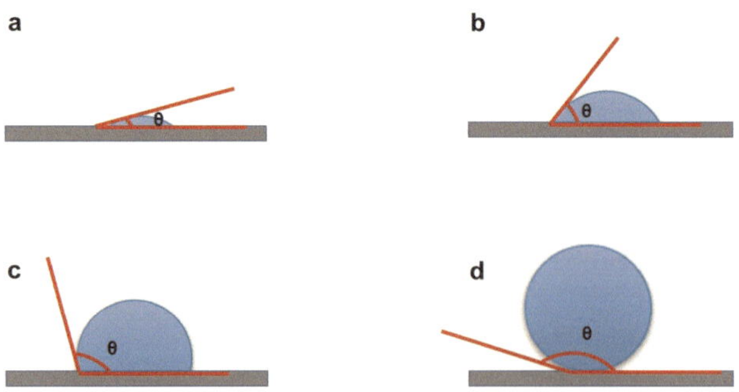

Figure 1.
Schematic representation of water contact angles for (a) super-hydrophilic, (b) hydrophilic, (c) hydrophobic, and (d) super-hydrophobic surfaces.

below 30°. However, the surface is considered as hydrophobic if the water contact angle is above 90°, and super-hydrophobic if the contact angle is higher than 150°.

The contact angle (θ) is a measure of the wettability of the surface and is defined as the angle formed between the solid-liquid interface and the tangent drawn at the liquid droplet at liquid-vapor interface. Contact angle hysteresis is the resistance against the movement of the water droplet along a solid surface. Super-hydrophobic surfaces exhibit water contact angles above 150° and contact angle hysteresis preferably lower than 10°. There are three models used to determine the water contact angle on any surface- Young's model, Wenzel's model and Cassie-Baxter's model [5].

2.1.1 Young's model of wetting

The Young's model is a very basic model for the measurement of the water contact angle and considers the substrate to be a flat and smooth surface, as shown in **Figure 2(a)**. At thermodynamic equilibrium conditions between the three phases i.e. solid, liquid and vapor phases, the water contact angle is determined using Young's equation (Eq. (1)).

$$Cos\ \theta = (\gamma_{SV} - \gamma_{SL})/\gamma_{LV} \tag{1}$$

where θ represents the contact angle of water droplet on the surface, γ_{SV} represents the surface energy at solid-vapor interface, γ_{SL} represents the surface energy at solid-liquid interface, and γ_{LV} represents the surface energy at liquid-vapor interface.

The Young's equation plays a crucial role in determining the wettability of the surface. The lower the contact angle formed, the higher is the surface's wettability. However, the wettability of the surface is also dependent on the surface roughness. Two models (i.e., Wenzel's and Cassie–Baxter's models) are mainly used to evaluate the water contact angle for the rough and chemically heterogeneous surfaces.

2.1.2 Wenzel's model of wetting

In the Wenzel's model, the substrate is a rough and heterogeneous surface, and the water droplet penetrates the surface cavities, as shown in **Figure 2(b)**. Due to the intimate contact of the water droplet with the microstructured surface, there is a change in the wetting behavior. The effect of roughness on the contact angle is represented by the Eq. (2).

$$Cos\ \theta_W = r\ Cos\ \theta \tag{2}$$

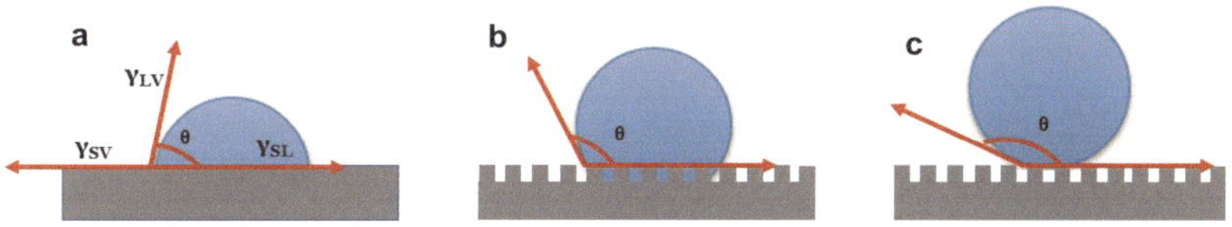

Figure 2.
Schematic representation of water contact angles for (a) Young's model, (b) Wenzel's model, and (c) Cassie–Baxter's model.

where θ_w is the Wenzel contact angle and r is the surface roughness factor, defined as the ratio of surface area of rough surface to the surface area of a flat projection of the same surface.

Since the surface roughness factor is greater than unity, the Wenzel contact angle increases for hydrophobic surfaces turning them into more hydrophobic and decreases for hydrophilic surfaces turning them into more hydrophilic surfaces. The Wenzel state demonstrates higher contact angle hysteresis (more than 10°) due to the penetration of the water droplet in the surface cavities.

2.1.3 Cassie-Baxter's model of wetting

Both the Wenzel and the Cassie–Baxter models explain the effect of surface roughness on the contact angle. In the Cassie–Baxter model, a heterogeneous wetting state is assumed in which air is entrapped in the surface cavities between water and the solid surface, as shown in **Figure 2**(c). Due to this air entrapment, water and solid interface area is reduced, and water and air interface area is increased. In such complex systems, the water contact angle is calculated as per the Cassie–Baxter equation, shown in Eq. (3).

$$Cos\ \theta_C = f_1\ Cos\ \theta_1 + f_2\ Cos\ \theta_2 \tag{3}$$

where θ_c is the Cassie–Baxter contact angle, f_1 and f_2 are the surface fractions of liquid-solid interface and liquid- vapor interface, respectively, and θ_1 and θ_2 are the contact angles of liquid-solid interface and liquid- vapor interface, respectively.

In the Cassie-Baxter state, as there is no penetration of the liquid in the surface cavities due to the presence of air pockets, lesser resistance is offered against the mobility of water droplets and thus, exhibit lower contact angle hysteresis (less than 10°). With the increase in the surface roughness of a hydrophobic substrate, the water contact behavior transits from the Wenzel model to the Cassie-Baxter model, leading to an increase in the contact angle and thus, changing the surface from hydrophobic to super-hydrophobic.

2.2 Super-hydrophobicity induced self-cleaning

Due to the higher water contact angles (above 150°) and lower contact angle hysteresis (less than 10°), super-hydrophobic coatings, instead of wetting the surface, clean themselves by rolling off water droplets that carry away any dirt. The mechanism of self-cleaning on a super-hydrophobic surface is shown in **Figure 3.**

The wetting behavior of a liquid on any solid surface is dependent on both the surface topography (physical roughness) and surface chemistry (surface energy).

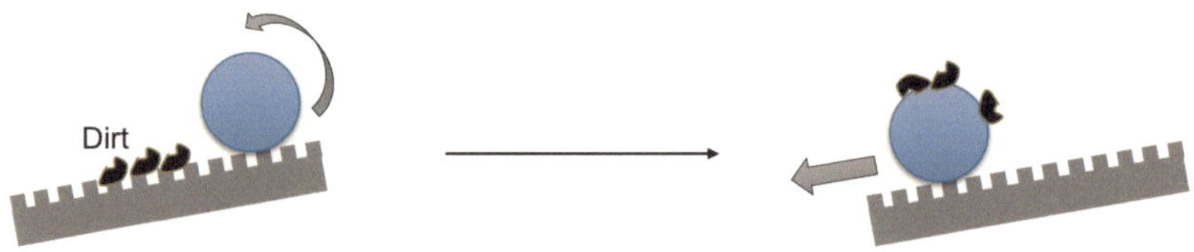

Figure 3.
Schematic representation for the mechanism of self-cleaning from a super-hydrophobic surface.

Based on these factors, the super-hydrophobicity on a surface can be achieved by majorly three different mechanisms [6]. The most common approach is the modification of surface topography by the incorporation of special nanostructures on the surface. The presence of nanostructures/nano-roughness on the surface leads to the formation of a lot of air gaps in between the water droplet and the surface, resulting in the surface morphology similar to Wenzel or Cassie-Baxter model, thus making the surface more hydrophobic. It also reduces the contact area between the surface and water droplet and thus, reducing the contact angle hysteresis. Due to these structures, the contact area between the solid surface and immobilized dirt particle is also reduced which reduces the adhesion force between the dirt particle and the surface, thus, enabling the easier removal of dirt by a drop of water rolling off the surface. The nanostructures-based roughness can be incorporated via different methods such as template-assisted method, photolithography, plasma treatment, electrospinning, chemical deposition method etc., as shown in **Figure 4** [7–9].

The second approach is based on the modification of surface chemistry to lower the surface free energy. When the surface energy of a solid surface is higher than that of a liquid, the liquid will tend to wet the surface of the solid to reduce the surface energy. However, when the surface energy of a solid surface is lower than that of a liquid, the liquid will not wet the surface and will try to achieve spherical shape to minimize the surface energy. Such a surface with lower surface energy is thus more hydrophobic than a surface with higher surface energy, as shown in **Figure 5**. This approach is achieved by the incorporation of low surface energy groups, such as fluoride and silicide, on a surface by grafting, spraying, or mixing.

The third approach utilizes the synergistic effect of both the above approaches. This could be achieved by incorporating low surface energy nanostructures on a solid surface, as shown in **Figure 6**.

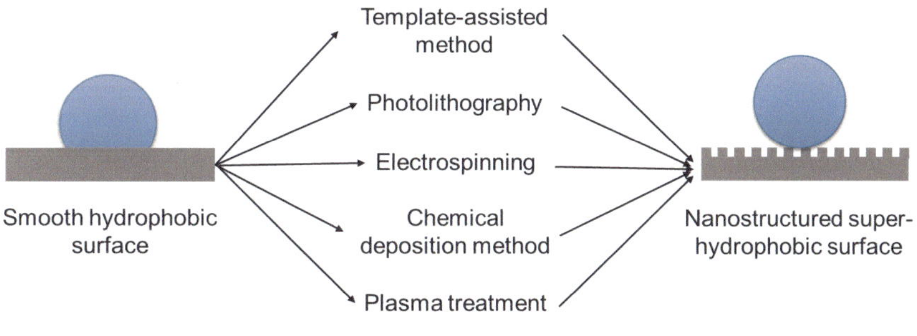

Figure 4.
Generation of nanostructured roughness on a surface for achieving super-hydrophobicity.

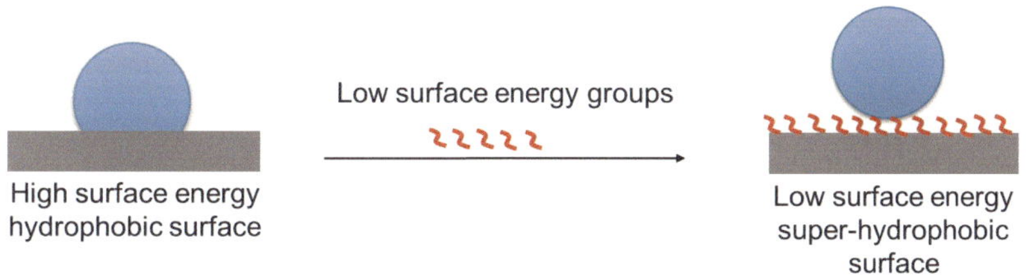

Figure 5.
Incorporation of low surface energy groups on a high surface energy surface for achieving super-hydrophobicity.

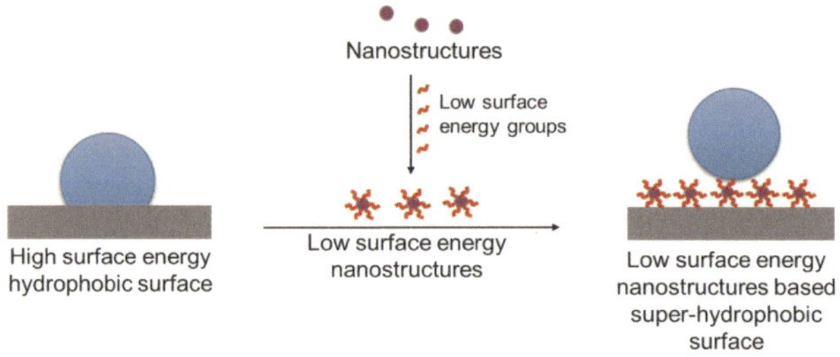

Figure 6.
Incorporation of low surface energy nanostructures on a surface for achieving super-hydrophobicity.

2.3 Super-hydrophilicity induced self-cleaning

Super-hydrophilic coatings exhibit higher surface energy on the solid surface and lower surface energy on liquid droplet. To lower the surface energy, water tends to wet the surface and spreads as a thin film rather than in the form of droplets. Such surfaces retain a thin film of water on the surface and thus, prevent the dirt from adhering to the surface. The dirt on the surface also gets readily washed away with the sheeting water layer, as shown in **Figure** 7 [10]. The super-hydrophilic coatings are generally developed by various methods such as sol-gel, layer by layer, chemical vapor deposition, physical vapor deposition, and spraying etc.

2.4 Photocatalysis induced self-cleaning

Photocatalysis is the most widely explored approach for self-cleaning coatings. During photocatalysis, an organic dirt or pollutant is degraded by a photocatalyst in the presence of sunlight. Various semiconductor nanoparticles such as TiO_2, ZnO, CuO, WO_3, and SnO_2 etc. have been explored as efficient photocatalysts for self-cleaning applications.

The semiconductor nanoparticles absorb radiations from the sunlight with energy equal to or greater than their band gap energy, leading to generation of charge carriers i.e. positively charged holes and negatively charged electrons. The electrons in the conduction band reduce the O_2 molecules into superoxide radical anion O_2^{-*} and the holes in the valence band oxidize H_2O into OH^*. The OH^* have an extremely high oxidation potential and can eventually degrade the dirt particles immobilized on the surface into carbon dioxide and water [11]. The mechanism of the photocatalytic degradation of dirt particles is shown in **Figure 8**. The photocatalytic process is majorly governed by the density of active species which is further controlled by the two competing processes of electron–hole pair generation and recombination.

TiO_2, being non-toxic, chemically inert and inexpensive semi-conducting material, has been significantly used for photocatalysis applications [12]. It exists in the form of

Figure 7.
Schematic representation for the mechanism of self-cleaning from a super-hydrophilic surface.

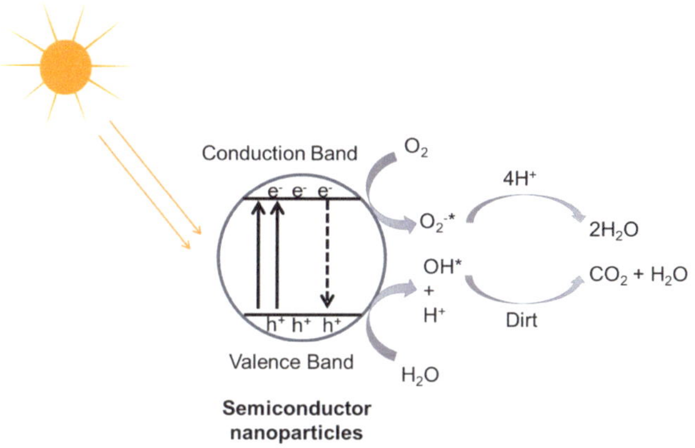

Figure 8.
Schematic representation of mechanism of photocatalysis.

three different crystalline structures, that is, anatase, rutile and brookite. Rutile is the most widely used form as it has a slightly higher refractive index. Both rutile and anatase are photoactive, with anatase form being more active due to a slower recombination rate of charge carriers. Brookite, being rarely found, is not used for any industrial applications. Since TiO_2 has a wide band-gap, the photocatalysis generally occurs in the presence of UV radiations and low energy visible light cannot be utilized to create the charge carriers. However, techniques such as ion-doping and heterostructure fabrication have been explored to achieve photocatalysis in the presence of visible light. Self-cleaning properties could further be improved by including strategies to increase the number and the lifetime of the generated electron-hole pairs.

3. Polyurethanes and their surfaces for self-cleaning

Polyurethane is a versatile polymer used in the coating industry. Polyurethane itself is a type of polymer that is connected to a chemical compound group known as carbamates. This polymer material is also thermosetting in nature; in other words, it burns rather than melts when heated. Another characteristic of polyurethane coatings is its customizability. A polyurethane coating is a polyurethane layer applied to the surface of a substrate for the purpose of protecting it or adding another value to it. These coatings help protect substrates from various types of defects such as corrosion, weathering, abrasion, and other deteriorating processes in addition to providing esthetics.

Based on the type of polyol used, there are three types of polyurethanes- polyester, polyether, and hybrid. The hardness and flexible performances of these coating layers can be controlled by formulation intelligence. These coatings can be formulated to be glossy, muted, opaque or transparent. While polyurethane coatings may appear to be visually similar to other coatings (e.g., epoxy), they possess several distinct properties that make them ideal for specific situations.

- While polyurethane coatings are relatively durable, they are softer and more elastic than their epoxy counterparts. This attribute makes polyurethane-coated floors ideal for moderate to heavy pedestrian traffic. The reduced stiffness gives polyurethane floors a slight elasticity, allowing them to absorb sharp impact loading.

- This durability also makes them more resistant to abrasion and less prone to dents and scratches.

- Improved elasticity also means that polyurethane floors can maintain their shape and mechanical properties in temperatures lower than 30°F (−1°C).

Due to its versatility, self-cleaning is an important value when it comes to polyurethane coatings. Achieving self-cleaning properties for polyurethane requires different formulation strategies as detailed in the Section 4.

4. Approaches to make polyurethane based self-cleaning surfaces

Self-cleaning coatings are primarily categorized into hydrophobic and hydrophilic types based on their behavior towards water. Hydrophobic coatings work on the principle of "lotus effect," inspired by the water-repellent properties of lotus leaves. Hydrophilic coatings typically contain metal oxides, such as titanium dioxide (TiO_2) or zinc oxide (ZnO), which have the ability to create a thin, continuous layer of water on the surface. In addition to the sheeting effect, metal oxides have an additional property of chemically breaking down complex dirt deposits by a sunlight assisted cleaning mechanism, that is, photocatalytic effect.

These processes are being leveraged by several researchers to successfully create artificial self-cleaning surfaces. For polyurethanes, hydrophilic/super-hydrophilicity, photocatalytic and lotus effect inspired super-hydrophobicity have been widely explored and are well documented [5].

4.1 Super-hydrophobic self-cleaning surfaces

Researchers have been inspired by the lotus leaf's ability to repel water and keep its surface clean, leading to extensive studies on the mechanistic aspects of this phenomenon [13, 14]. The surface of a lotus leaf is covered with tiny microscale and nanoscale structures, which give it a rough and textured surface. These structures, often referred to as papillae and epicuticular wax crystals, create a hierarchical roughness that minimizes the contact area between the leaf and water droplets. As a result, the droplets rest on the surface with minimal contact, allowing them to roll off easily (**Figure 9**) [15]. With the mechanistic understanding, the concept is well explored by several researchers to create artificial super-hydrophobic surfaces whose water contact angle (WCA) is greater than 150° [16]. The latest strategies to prepare super-hydrophobic polyurethanes (SHPU) are discussed in below section:

4.1.1 Incorporation of low surface energy substance

Surface energy plays a crucial role in determining the wettability of solid surfaces. Silicide and fluoride materials are commonly used for their low surface energy properties in order to create hydrophobic surfaces. These materials can be incorporated into solid surfaces through various processes such as grafting, spraying, or mixing.

Figure 9.
Lotus leaves with super-hydrophobicity.

4.1.1.1 Incorporation of silicide

Silicide possesses low surface energy and its incorporation into the solid can lead to the enrichment of silicon on the surface of PU, thereby decreasing the surface energy and increasing the hydrophobicity. Nano-SiO_2 is the commonly used silicide to improve the hydrophobicity of PU. It can be incorporated onto the surface of the PU substrate by spraying a nano-SiO_2 dispersion to form a coating layer with low surface energy [17]. A study created a SHPU film by spraying a PU/SiO_2/chloroform dispersion onto a PU sponge. The resultant SHPU sponge was able to separate oil from water in an oil in-water emulsion via absorption [18]. SHPU can also be prepared by spraying nano-SiO_2/PU dispersion thus incorporating both on the surface and into the bulk of PU, which is beneficial in retaining the super-hydrophobicity even when the surface of PU is damaged.

Surface functionalization of nano-SiO_2 has also been found to enhance their interaction with PU surface. Furthermore, the functionalization of nano-SiO_2 with long-chain silane has shown to be able to further increase the water contact angle (WCA) value of PU.

4.1.1.2 Incorporation of fluoride

Fluoride substances contribute to low free energy and are widely explored in the preparation of SHPU [19]. The surface energies of various substances containing C–F bonds and/or -CF_3 groups is well summarized in literature [6]. When incorporating fluoride into PU, nano-SiO_2 can be used as a "connector" or coupling agent to facilitate the interaction between the PU surface and the fluoride compounds (**Figure 10**). The incorporation of nano-SiO_2 into PU resulted in a maximum WCA of

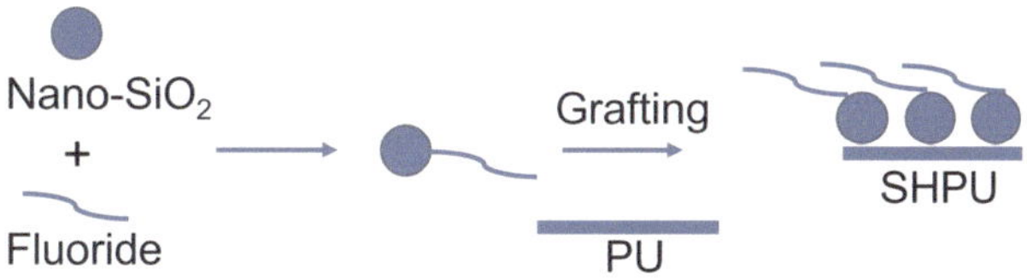

Figure 10.
Grafting of fluoride incorporated nano-SiO_2 onto PU.

160°, however, the fluoride-modified nano-SiO_2 was able to give a maximum WCA of about 170°. A durable and superhydrophobic waterborne PU (SHWPU) coating was prepared by spraying waterborne PU (WPU) on a panel and then fluoride modified nano-SiO_2 was subsequently sprayed on it [20, 21]. In addition, a perfluoroalkyl methacrylic copolymer was used to prepare superhydrophobic polyurethane.

However, currently, usage of fluorine is not considered to be environment friendly, hence, the fabrication of superhydrophobic surfaces without fluorine could become one of the main strategies in the future. In this regard, PDMS has recently been employed by few groups to introduce self-cleaning properties.

4.1.2 Construction of rough surfaces on PU

4.1.2.1 Electrospinning process

Electrospinning is a technique used to prepare nanofibers and is widely explored to construct rough surfaces on PU and several other natural and synthetic polymers. During the process, the polymer solution is stretched into nanofibers under the influence of an electric field. After that, these nanofibers are collected to form a rough film with a nanofiber network, thereby enhancing the hydrophobicity of the film [22].

4.1.2.2 Incorporation of nanoparticles

Incorporating nanoparticles can modify the surface roughness. Currently, nanoparticles such as carbon nanoparticles, that is, carbon nanotubes (CNT) and carbon nanofibers (CNF), and metal oxides, that is, molybdenum disulfide (MoS_2), titanium dioxide (TiO_2), aluminum oxide (Al_2O_3), and zinc oxide (ZnO) have been used to furnish rough surfaces.

4.1.3 Combined strategy of surface energy reduction and surface roughness generation

The combined strategy of surface energy reduction and surface roughness generation has revealed the best results for attaining super-hydrophobicity in comparison to using either of two approaches alone. Using this strategy, some of the common technical methods to achieve super-hydrophobicity are by grafting silane modified nanoparticles onto PU substrate, grafting fluoride onto rough substrate, or by grafting graphene onto PU surface [23].

Organic fluorine and silicone are commonly used as modifiers to enhance hydrophobicity, but they can have limitations such as poor low-temperature resistance and toxicity. To address these limitations and simplify the modification process, a biomimetically hierarchical structure has been reported for achieving superhydrophobic and self-cleaning properties in thermoplastic polyurethane (TPU) surfaces. This approach involves electrospinning TPU fibers and electrospraying TPU microspheres [24]. During the preparation of microspheres, environmentally friendly, fluorine-free hexadecyl trimethoxysilane (HDTMS) is added into the electrospray solution, to reduce surface energy. This approach offers advantages such as using environmentally friendly materials, avoiding toxicity concerns associated with fluorine-modified compounds, and simplifying the modification process.
Another environmentally benign self-cleaning amorphous-SiO_2/nano-TiO_2 based

HDTMS coating with super-hydrophobicity and photocatalytic activity was also studied and reported recently [25].

4.2 Super-hydrophilic (photocatalytic) self-cleaning surfaces

The combination of photocatalysis and water sheeting on hydrophilic surfaces offers another efficient self-cleaning mechanism. The photocatalytic activity of the surface breaks down dirt, while the super-hydrophilic nature (contact angle $<30°$), ensures that water spreads evenly and efficiently to wash away the debris. TiO_2 is one of most widely explored materials for self-cleaning. Due to its special photo-induced properties—photocatalysis and photo-induced super-hydrophilicity - a wide variety of TiO_2-based super-hydrophilic self-cleaning surfaces have been fabricated using different synthetic approaches on different substrates. The photocatalytic performance of TiO_2 is researched to be improved by doping with metals with higher oxidation states. Most of the studies explored using phase separated dopants with the incorporation of nanoparticles. Such phenomenon is also well depicted and published for polyurethanes.

Recent studies showed an introduction of hollow nano-TiO_2 spheres (HNTSs) with high photocatalytic and permeability performances as well as introducing large surface area into the PU film to endue its superior water vapor permeability property, water resistance along with desirable self-cleaning performance [26].

Hydrophilic surfaces prepared by silicon dioxide have also attracted much attention. Studies are also available wherein superhydrophilicity attained by silica nanoparticle film exhibits self-cleaning property without using any photocatalytic materials.

Studies also showed the improved self-cleaning property by using silicon dioxide sol modifying the wettability of acrylate polyurethane from hydrophobicity to hydrophilicity. Transparent hydrophilic photocatalytic TiO_2/SiO_2 thin films were also explored in polycarbonate substrate which was precoated by an intermediate SiO_2 layer. The coated surfaces displayed considerable photocatalytic activity and superhydrophilicity. The self- cleaning coatings usually suffer from photodegradation caused by the nanofiller's photoactivity. From the previous study, it was found that coating of SiO_2 on the surface of TiO_2 nanoparticles influenced the photocatalytic activity of the formed PU composite films, resulting in reduced photodegradation. Recent studies report the integration of SiO_2-coated TiO_2 nanostructures into a PU matrix for the synthesis of self-cleaning coatings [27].

Photochemical additives based on C60 fullerene were incorporated into polyurethane coatings to investigate their coating compatibility and ability to impart chemical decontaminating capability to the coating surface. C60 fullerene molecules have also been observed to exhibit intriguing photochemical properties, including oxidative capabilities, which hold exciting potential for development of a self-decontaminating coating [28].

A combination of easy cleaning with self-cleaning of oleophobic and hydrophilic stains has been recently demonstrated in PU system by using a combination approach of hydrophilic thermo-responsive hydrogel coating and the self-cleaning from the embedded non-metallic photocatalyst g-C_3N_4 [29]. Due to the existence of strong hydrogen bonds between the hydroxyl groups in the hybrid hydrogel coating and the hydroxyl/carboxyl groups in the plasma-treated PU, the hybrid hydrogel coating is very stable on PU. Simultaneously, the acrylamide network in the hybrid hydrogel coating enhances its mechanical strength. Because the transition temperature of oligo(ethylene glycol) methyl ether methacrylate ($OEGMA_{300}$) is well above the room temperature, the cross-linked coating remains hydrophilic in ambient conditions. Thus, oleophilic stains, such as oil and grease, can be easily removed from the coating surface. In addition, the embedded photocatalyst g-C_3N_4 in the hybrid hydrogel coating introduces the extra capability of decomposing organic compounds under sunshine, which favors the removal of hydrophilic stains such as dyes and wines.

4.3 Oleophobic surfaces for self-cleaning

Oleophobic surfaces play a crucial role in self-cleaning and anti-fouling applications, particularly in industries such as oil, steel, and marine environments where oil spills can cause significant damage [30]. Super-oleophobic surfaces, which exhibit strong repellency towards oil and organic liquids with lower surface tension, have garnered interest for their potential in preventing oil adhesion and facilitating easy cleaning. To create super-oleophobic surfaces, one approach is to design solid surfaces with a lower surface energy than that of oil. By reducing the surface energy of the material below the surface tension of the oil, the oil droplets tend to minimize contact with the surface, leading to oil repellency. Hydrophobic and oleophobic coatings can be formulated using ceramic particles such as SiO_2 (silicon dioxide), SiO (silicon monoxide), and Al_2O_3 (aluminum oxide). These ceramic particles offer thermal and chemical durability, making them suitable for various applications. Such coatings serve as alternatives to materials like Teflon, which is known for its oil-repellent properties but may have limitations in certain environments. Super-oleophobic coatings are also developed by employing techniques like spray casting of nanoparticle-polymer suspensions. This process allows for the deposition of a thin film containing ceramic nanoparticles onto a substrate, creating a surface with excellent oil-repellent properties [31].

4.4 Amphiphobic surfaces for self-cleaning

Amphiphobic coatings repel both water and oil, hence surface combines both hydrophobicity and oleophobicity. Inspired by nature, researchers have fabricated superhydrophobic-superoleophobic surfaces. A stable superamphiphobic coating using electrospinning technique was fabricated on glass substrate with nano/mesostructured TiO_2 [32]. Also, a novel hydrophobic and oleophobic surfaces using polyurethane with hydrogenated polyisoprene soft segment have been fabricated and reported recently [33].

5. Applications of polyurethane self-cleaning coatings

Polyurethane based self-cleaning coatings are popular in different applications such as antifogging, anti-icing, anti-reflection, corrosion resistance, drag reduction, sensors, solar cells, and textiles etc. Potential application sectors include but not limited to textile industry (self-cleaning clothing), automobile industry (self-cleaning windshield glass, car bodies and mirrors), optical industry (cameras, sensors, lenses and telescopes), marine industry (anticorrosion protection) and aerospace industry (non-sticky surfaces) and so on. Moreover, self-cleaning coatings can also be used in windows (window coatings), solar modules (self-cleaning coatings for solar modules) and in paints (exterior paints with self-cleaning properties) etc. Commonly desired properties for such applications include durability, water and dirt repellency, easy cleaning, scratch resistance, chemical resistance, UV resistance, etc.

5.1 Anti-fogging

Polyurethane-based self-cleaning coatings can be designed to provide an anti-fogging effect, which can be particularly useful in applications where visibility is important. This is achieved through the hydrophobic properties of the coating, which prevent moisture from condensing on the surface. When a surface is cooled below the dew point, moisture in the air can condense on the surface, leading to the formation of fog. Interestingly, a hydrophobic polyurethane coating can prevent this from happening by repelling water droplets and preventing them from coalescing into larger droplets that can cause fogging. By preventing fogging, these coatings can help to improve safety and performance, as well as reduce the need for frequent cleaning.

Along with hydrophobicity, the polyurethane formulations for self-cleaning coatings may also contain anti-fogging agents that help to further reduce the formation of fog. These agents work by lowering the surface tension of the water droplets, causing them to spread out and evaporate more quickly. This type of polyurethane-based self-cleaning anti-fog coatings can be particularly beneficial in applications such as automotive windshields, eyewear, and medical equipment where visibility is crucial.

A steel surface coated using polyurethane, SiO_2 nanoparticles and hexadecyltrimethoxysilane by a spin-coating technique has been reported to show excellent superhydrophobic and anti-fogging effect [34]. Highly transparent antifogging polyurethane coatings have been fabricated via a UV-assisted cross-linking method by Li et al. [35].

5.2 Anti-icing

In cold regions, especially during winter season, layers of ice get deposited on solid materials exposed to open environments. Polyurethane-based self-cleaning coatings when applied to surfaces, the coatings create a water-repellent layer due to the hydrophobic nature that prevents water from adhering to the surface. This can help to reduce the formation of ice significantly well [36]. Self-cleaning polyurethane superhydrophobic coating having excellent anti-icing property at $-10°C$ could be a viable strategy for transmission line and wind turbines icing mitigation [37].

5.3 Anti-reflective

The anti-reflective effect of polyurethane-based self-cleaning coatings can be particularly beneficial in applications such as eyewear, camera lenses, and electronic displays, where clarity and visibility are important. By reducing glare and reflection, these coatings can improve the performance and usability of these devices. Polyurethane-based self-cleaning coatings can also provide an anti-reflective effect, which can improve visibility and reduce glare. This is achieved through the use of special coatings that are designed to reduce the reflection of light on the surface. When light hits a surface, some of it is absorbed while the rest is reflected back. This reflection can cause glare and reduce visibility, particularly in bright sunlight or under certain lighting conditions. However, an anti-reflective coating can reduce the amount of light that is reflected back, allowing more light to pass through the surface and improving visibility. In addition, some formulations of self-cleaning polyurethane coatings may also contain anti-reflective agents that further reduce the reflection of light. These agents work by altering the refractive index of the coating, causing light to be refracted in a way that reduces the amount of reflection. A polyurethane coating with photocatalytic properties can help to reduce the accumulation of ice and snow on the coated surface. Ko et al. [38] have reported a replication route to non-planar, three dimensional microlens arrays with an antireflective poly(urethane) surface with "moth-eye" nanopattern.

5.4 Enhanced corrosion resistance coatings

Polyurethane can directly adhere to metals without the use of adhesion promotors. Hence, polyurethanes are used in direct-to-metal coatings. Polyurethane-based self-cleaning coatings can provide an effective barrier against corrosion and protect metal surfaces from environmental damage. The hydrophobic and/or photocatalytic properties of these coatings prevent the accumulation of moisture and contaminants on the surface, which can potentially cause corrosion. The formulations of self-cleaning polyurethane coatings can be designed to contain corrosion inhibitors that provide additional protection against corrosion. These inhibitors work by forming a protective layer on the surface of the metal, preventing corrosive agents from penetrating the surface. The corrosion resistance effect of polyurethane-based self-cleaning coatings can be particularly useful in applications such as automotive, marine, and aerospace industries where metal components are exposed to very harsh environments. By protecting these components from corrosion, these coatings can help to extend the lifespan of the equipment and reduce maintenance costs. Nosrati et al. [39] have prepared anti-corrosive polyurethane coating modified with titanium dioxide/polyaniline/halloysite nanotube/carbon nanotube nanocomposite having excellent effect to improve environmental protection. Ye et al. [40] reported that ZnO/polyurethane nanocomposite coatings can enhance the corrosion resistance of stainless-steel. Zhang et al. [41] fabricated slippery coating by spraying silicone-oil soaked SiO_2 with irregular coral cluster structure on the acrylic polyurethane having corrosion resistance with self-cleaning and good coating stability. Fluorine-modified hyperbranched waterborne polyurethane resin applied to a water-based nano anti-corrosion self-cleaning finish paint has been reported to show excellent anti-corrosion property [42].

5.5 Drag reduction

Frictional drag is a major source of resistance that affects the speed and efficiency of vehicles and equipment. By reducing friction, a polyurethane-based self-cleaning coating can improve the flow of fluids around the surface, reducing drag and increasing speed and efficiency. Polyurethane-based self-cleaning coatings can also provide a drag reduction effect, which can improve the performance of vehicles and equipment that move through fluids, such as air or water. The hydrophobic properties of these coatings reduce the surface tension of the fluid, allowing it to slide more easily over the surface and reducing friction. In addition, some formulations of self-cleaning polyurethane coatings may also contain additives that further reduce drag, such as nanoparticles that modify the flow of the fluid around the surface.

The drag reduction effect of polyurethane-based self-cleaning coatings can be particularly beneficial in applications such as aerospace, marine, and automotive industries, where reducing drag can improve fuel efficiency and reduce operating cost. By improving the performance of these systems, these coatings can help to reduce environmental impact and improve sustainability. Superhydrophobic PDMS assisted immobilized fluorine functionalized SiO_2 nanoparticles have been reported to improve coating adhesion to substrate materials [43].

5.6 Sensors

Polyurethane-based self-cleaning coatings can be applied in various sensor applications to improve their performance and longevity. In optical sensors, the anti-reflective properties of the coating can reduce the reflection of light and improve the accuracy of the sensor. Similarly, in gas sensors, the hydrophobic and/or photocatalytic properties of the coating can prevent the accumulation of moisture and contaminants on the surface of the sensor, reducing interference with the sensor's performance [44]. The anti-fouling properties of polyurethane-based self-cleaning coatings can help to prevent the buildup of biofilms and other materials on the surface of sensors used in medical and biological applications. A group of researchers fabricated a pressure sensor consisting of PU mesoscaled dome arrays embedded with gradient-distributed silver nanowire (AgNW) for flexible electronic applications [45].

5.7 Solar

Polyurethane-based self-cleaning coatings can be applied to solar cells to improve their efficiency & reduce maintenance requirements. The coatings can provide several benefits, including self-cleaning, anti-reflective, anti-scratch, hydrophobic and UV-stability. Interpenetrated polymer networks prepared using polyurethane- acrylic colloidal suspension when applied on glass covers of solar cell panels is reported to give excellent super-hydrophobicity, transparency, and durability [46].

5.8 Textile

Textiles coated with these types of coatings can resist dirt, dust, and other contaminants, making them easier to clean and maintain.

The hydrophobic properties of polyurethane-based coatings can prevent the absorption of moisture and liquids, making textiles more resistant to water damage.

Aliphatic Polyurethane-based coatings can provide protection against UV radiation, which can cause damage to textiles and fade their colors over time.

The textiles coated with these types of coatings can be more durable, with improved resistance to wear and tear, stretching, and abrasion. Chen et al. [47] fabricated coating with self-cleaning and pH-controllable oil/water separating ability on account of the pH-responsive UV-cured PU having good super-hydrophobicity and can maintain their self-cleaning ability even after mechanically damaged, seawater immersing and UV irradiation.

5.9 Others: electrowetting and other functions

Electrowetting is a phenomenon that occurs when an electric field is applied to a liquid droplet on a surface, causing the droplet to spread or contract. Incidentally, polyurethane based self-cleaning coatings can exhibit electrowetting behavior, which can be beneficial in certain applications. By applying an electric field to the surface of a polyurethane based self-cleaning coating, the contact angle between the droplet and the surface can be modified. This can change the wetting behavior of the droplet and allow for precise control of its movement and positioning on the surface. This behavior can be useful in microfluidic systems and lab-on-a-chip devices, where precise control of droplet movement is important.

6. Current challenges and possible solutions

Along with versatility of polyurethane self-cleaning coatings, there are still several challenges that must be addressed to develop multi-functional robust and effective self-cleaning coatings to take the next level. The current challenges and possible solutions are highlighted below:

6.1 Durability

One of the main challenges with aromatic polyurethane based self-cleaning coatings is maintaining their durability over time, particularly in the UV light-based environments. Possible solutions to this challenge include incorporating additives that enhance the coating's UV resistance, abrasion and scratch resistance or modifying the coating's structure to improve its UV resistance and adhesion and cohesion properties.

6.2 Compatibility with different substrates

Although polyurethane is versatile, another challenge is developing coatings that can be applied to a wide range of substrates, including metals, plastics, and ceramics. Possible solutions to this challenge include optimizing the coating's composition and application process to improve its adhesion to different substrates, or developing surface pre-treatments that improve the substrate's surface energy and adhesion properties.

6.3 Environmental impact

Many current self-cleaning coatings contain toxic chemicals or are not biodegradable, which can have negative environmental impacts. Possible solutions to this challenge include using eco-friendly raw materials and production processes, or developing coatings that are biodegradable or recyclable.

6.4 Cost

Polyurethane based self-cleaning coatings can be more expensive than traditional coatings, which can limit their adoption in some applications. Possible solutions to this challenge include developing more cost-effective production processes, or identifying applications where the benefits of self-cleaning coatings outweigh the higher cost.

6.5 Performance under extreme conditions

Self-cleaning coatings may not perform as well under extreme conditions, such as high temperatures (>200°C), humidity, or exposure to chemicals. Possible solutions to this challenge include developing coatings with improved thermal stability, resistance to chemical exposure, or moisture resistance.

Overall, developing a robust, high performance and cost-effective polyurethane based self-cleaning coatings requires a combination of material chemistry, and engineering. In order to provide a solution-based approach to these challenges, researchers can continue to improve the performance and applicability of self-cleaning coatings in a wide range of industries and applications.

7. Characterization of polyurethane self-cleaning surfaces

The self-cleaning property of a polyurethane surface is mainly governed by its *phobic* and *philic* nature towards oil and water. The exact characterization is made by quantitatively measuring (1) surface energy, (2) surface roughness and (3) liquid droplet contact angle by using various characterization techniques and correlate this with the self-cleaning effect. Typically, higher surface energy polymers offer to be potential adhesives since, higher their surface free energy higher will be their ability to interact with the substrate. The surface energy aspects can be simulated with different mathematical models to design new surfaces.

8. Conclusions

In conclusion, all the research efforts, as we have seen here, are explored to mimic the supreme strategies perfected by nature over billions of years. The self-cleaning surface on naturally occurring leaves and wings of certain insects is multipurpose in achieving self-cleaning, anti-reflective, camouflage and various other functionalities which has got researchers across the globe take stock and attempt to mimic. Though the self-cleaning surfaces designed by them are yet to match their naturally occurring counterparts, the fabrication techniques have indeed evolved into more environmentally compatible and cost-effective and polyurethane is one amongst them.

Polyurethane is a well-accepted polymer-based coating material in the market. Polyurethane polymers can be tailored very easily to meet the *phobic* and *philic* nature to oils and water. Therefore, polyurethane-based self-cleaning coatings have potential for reducing maintenance and cleaning costs in various industries. Polyurethane coatings can be tuned to utilize hydrophobic and/or photocatalytic properties to repel dirt, dust, and other contaminants, which can then be easily washed away as exemplified in this chapter. Polyurethane based applications for self-cleaning coatings are currently in automotive, aerospace, and marine industries.

As the market expands more effort on building a strong structure performance analysis is critical to develop a robust self-cleaning surface for different applications. The future outlook for self-cleaning polyurethane coatings is promising, with potential for further improvements in the smartness, durability, efficiency, sustainability/eco-friendliness and circularity of the coatings along with the development of easily scalable manufacturing processes. As and when more industries adopt these coatings, there may be an increased demand for customized formulations to meet specific application requirements in this domain.

Acknowledgements

The authors would like to acknowledge Rob Kerkhofs for his contribution to providing literature support.

Conflict of interest

The authors declare no conflict of interest.

Author details

Kamlesh Panwar, Neelam Vyas, Roopali Rai* and Shreedhar Bhat*
Huntsman Solutions India Pvt. Ltd., Mumbai, India

*Address all correspondence to: roopali_rai@huntsman.com and shreedhar_bhat@huntsman.com; shreedharbhat@gmail.com

References

[1] Das A, Mahanwar P. A brief discussion on advances in polyurethane applications. Advanced Industrial and Engineering Polymer Research. 2020; **3**(3):93-101. DOI: 10.1016/j.aiepr.2020.07.002

[2] Bayer O. Das di-lsocyanat-polyadditionsverfahren (polyurethane). Angewandte Chemie. 1947;**59**:257-272. DOI: 10.1002/ANGE.19470590901

[3] Akindoyo JO, Beg MDH, Ghazali S, Islam MR, Jeyaratnam N, Yuvaraj A. Polyurethane types, synthesis and applications- a review. RSC Advances. 2016;**6**:114453-114482. DOI: 10.1039/C6RA14525F

[4] Paz Y, Luo Z, Rebenberg L, Heller A. Photooxidative self-cleaning transparent titanium dioxide films on glass. Journal of Materials Research. 1995;**10**: 2842-2848. DOI: 10.1557/JMR.1995.2842

[5] Padmanabhan NT, John H. Titanium dioxide based self-cleaning smart surfaces: A short review. Journal of Environmental Chemical Engineering. 2020;**8**(5):104211. DOI: 10.1016/j.jece.2020.104211

[6] Zhao H, Gao WC, Li Q, Khan MR, Hu GH, Liu Y, et al. Recent advances in superhydrophobic polyurethane: Preparations and applications. Advances in Colloid and Interface Science. 2022; **303**:102644. DOI: 10.1016/j.cis.2022.102644

[7] Sethi SK, Manik G. Recent progress in super hydrophobic/hydrophilic self-cleaning surfaces for various industrial applications: A review. Polymer-Plastics Technology and Engineering. 2018; **57**(18):1932-1952. DOI: 10.1080/03602559.2018.1447128

[8] Konar M, Roy B, Govindaraju T. Molecular architectonics-guided fabrication of superhydrophobic and self-cleaning materials. Advanced Materials Interfaces. 2020;**7**(11): 2000246. DOI: 10.1002/admi.202000246

[9] Wei X, Niu X. Recent advances in superhydrophobic surfaces and applications on wood. Polymers. 2023;**15**(7):1682. DOI: 10.3390/polym15071682

[10] Otitoju TA, Ahmad AL, Ooi BS. Superhydrophilic (superwetting) surfaces: A review on fabrication and application. Journal of Industrial and Engineering Chemistry. 2017;**47**:19-40. DOI: 10.1016/j.jiec.2016.12.016

[11] Panwar K, Jassal M, Agrawal AK. TiO_2–SiO_2 Janus particles with highly enhanced photocatalytic activity. RSC Advances. 2016;**6**:92754-92764. DOI: 10.1039/C6RA12378C

[12] Panwar K, Jassal M, Agrawal AK. TiO_2–SiO_2 Janus particles for photocatalytic self-cleaning of cotton fabric. Cellulose. 2018;**25**:2711-2720. DOI: 10.1007/s10570-018-1698-2

[13] Liu K, Jiang L. Bio-inspired self-cleaning surfaces. Annual Review of Materials Research. 2012;**42**:231-263. DOI: 10.1146/annurev-matsci-070511-155046

[14] Ganesh VA, Raut HK, Nair AS, Ramakrishna S. A review on self-healing coating. Journal of Materials Chemistry. 2011;**21**(41):16304-16322. DOI: 10.1039/C1JM12523K

[15] Prathapan R, Ganesh VA, Nair SV, Nair AS. A review on self-cleaning and

multifunctional materials. Journal of Materials Chemistry A. 2014;**2**: 14773-14797. DOI: 10.1039/c4ta02542c

[16] Dejun M, Yingmo C, Jiachen J, Lingxiao K, Yuming X. Preparation Method for Environmentally Friendly Super-Hydrophobic Coating with High Mechanical Durability. China: Univ Shandong Science & Tech; 2020 WO2020186579A1

[17] Wu Z, Wang H, Xue M, Tian X, Ye X, Zhou H, et al. Facile preparation of superhydrophobic surfaces with enhanced releasing negative air ions by a simple spraying method. Composites Science and Technology. 2014;**94**: 111-116. DOI: 10.1016/j.compscitech.2014.01.019

[18] Barroso-Solares S, Pinto J, Fragouli D, Athanassiou A. Facile oil removal from water-in-oil stable emulsions using PU foams. Materials (Basel). 2018;**11**(12):2382. DOI: 10.3390/ma11122382

[19] Qiuyu Z, Kang F, Hepeng Z, Baoliang Z. Fluorinated Polyurethane Coating and Preparation Method of Super-Hydrophobic Coating. China: Univ Northwestern Polytechnical; 2020 CN110845952A

[20] Zheng H, Pan M, Wen J, Yuan J, Zhu L, Yu H. Robust, transparent and superhydrophobic coating fabricated with waterborne polyurethane and inorganic nanoparticle composites. Industrial and Engineering Chemistry Research. 2019;**58**(19):8050-8060. DOI: 10.1021/acs.iecr.9b00052

[21] Ruilin L, Huafeng L, Jijun F. Waterborne Durable Super-Hydrophobic Organic Silicon Emulsion Material and Preparation Method Thereof. China: Nanjing Hunriss New Material Tech Co Ltd; 2022 CN114395327A

[22] Yang X, Guo Y, Han Y, Li Y, Ma T, Chen M, et al. Significant improvement of thermal conductivities for BNNS/PVA composite films via electrospinning followed by hot-pressing technology. Composites Part B: Engineering. 2019; **175**(31):107070. DOI: 10.1016/j.compositesb.2019.107070

[23] Xuedong S, Liqiang H. Super-Hydrophobic Graphene Coating as Well as Preparation Method and Application Thereof. China: Shenzhen Graphene Land Legend Tech Co Ltd.; 2021 CN113025187A

[24] Lin L, Choi Y, Chen T, Kim H, Lee KS, Kang J, et al. Superhydrophobic and wearable TPU based nanofiber strain sensor with outstanding sensitivity for high-quality body motion monitoring. Chemical Engineering Journal. 2021;**419**:129513. DOI: 10.1016/j.cej.2021.129513

[25] Wang X, Ding H, Wang C, Zhou R, Li Y, Li W, et al. Self-healing superhydrophobic A-SiO_2/N-TiO_2@HDTMS coating with self-cleaning property. Applied Surface Science. 2021;**567**:150808. DOI: 10.1016/j.apsusc.2021.150808

[26] Guo X, Lyu B, Gao D, Cheng K, Ma J, Yang N, et al. Water vapor permeability and self-cleaning properties of solventree polyurethane improved by hollow nano-TiO_2 spheres. Journal of Applied Polymer Science. 2021;**138**(1):49638. DOI: 10.1002/app.49638

[27] Chen C, Xu WZ, Charpentier PA. SiO_2 encapsulated TiO_2 nanotubes and nanofibers for self-cleaning polyurethane coatings. Journal of Photochemistry and Photobiology A: Chemistry. 2017;**348**:226-237. DOI: 10.1016/j.jphotochem.2017.08.014

[28] Lundin JG, Giles SL, Cozzens RF, Wynne JH. Self-cleaning photocatalytic polyurethane coatings containing modified C60 fullerene additives. Coatings. 2014;**4**(3):614-629. DOI: 10.3390/coatings4030614

[29] Fang Z, Zhang J, Yan X, Hu L, Lei L, Fan H, et al. Simultaneous and efficient removal of oleophilic and hydrophilic stains from polyurethane by the combination of easy-cleaning and self-cleaning. ACS Applied Materials & Interfaces. 2022;**14**(14):16641-16648. DOI: 10.1021/acsami.2c01042

[30] Michael HF. Flexible Superhydrophobic and/or Oleophobic Polyurethane Coatings. USA: Ross Tech Corporation; 2019 US20190315973A1

[31] Cheng DF, Urata C, Yagihashi M, Hozumi A. A statically oleophilic but dynamically oleophobic smooth non-perfluorinated surface. Angewandte Chemie. 2012;**51**(12):2956-2959. DOI: 10.1002/anie.201108800

[32] Yong J, Chen F, Yang Q, Huo J, Hou X. Superoleophobic surfaces. Chemical Society Reviews. 2017;**46**(14): 4168-4217. DOI: 10.1039/C6CS00751A

[33] Shiraki Y, Yokoyama H. Novel hydrophobic and oleophobic surfaces using polyurethane with hydrogenated polyisoprene soft segment. Materials Today Communications. 2020;**24**: 101243. DOI: 10.1016/j.mtcomm.2020.101243

[34] Meena MK, Tudu BK, Kumar A, Bhushan B. Development of polyurethane-based superhydrophobic coatings on steel surfaces. Philosophical Transactions A. 2020;**378**:20190446. DOI: 10.1098/rsta.2019.0446

[35] Li T, Wang Y, Liu X, Yang X, Zhang S, Wang S, et al. Transparent UV-cured polyurethane–surfactant coatings with antifogging properties. ACS Applied Polymer Materials. 2022; **4**(11):8147-8155. DOI: 10.1021/acsapm.2c01092

[36] Qiuyu Z, Yi T, Hepeng Z, Baoliang Z. Method for Preparing Self-Cleaning Anti-Icing Coating Based on Brushlike. China: Univ Northwestern Polytechnical; 2022 US20220154036A1

[37] Lv L, Liu H, Zhang W, Chen J, Liu Z. Facile UV-curable fabrication of robust, anti-icing superhydrophobic coatings based on polyurethane. Materials Letters. 2020;**258**:126653. DOI: 10.1016/j.matlet.2019.126653

[38] Ko DH, Tumbleston JR, Henderson KJ, Euliss LE, DeSimone JM, Lopez R, et al. Biomimetic microlens array with antireflective "moth-eye" surface. Soft Matter. 2011;**7**(14): 6404-6407. DOI: 10.1039/C1SM05302G

[39] Nosrati R, Kiani G, Ghavidel AK, Rashidi A. Improving environmental protection of waterborne polyurethane coating by adding TiO_2/polyaniline/HNT/CNT nanocomposite. Environmental Science and Pollution Research International. 2020;**27**(6): 6438-6448. DOI: 10.1007/s11356-019-07333-x

[40] Ye X, Wang Z, Ma L, Wang Q, Chu A. Zinc oxide array/polyurethane nanocomposite coating: Fabrication, characterization and corrosion resistance. Surface and Coatings Technology. 2019;**358**:497-504. DOI: 10.1016/j.surfcoat.2018.11.080

[41] Zhang K, Xu S, Tan J, Li X, Du L, Ma J, et al. Fabrication of liquid-infused porous slippery coating with stability, self-cleaning and corrosion resistance properties for carbon steel. Colloids and Surfaces A: Physicochemical and

Engineering Aspects. 2023;**669**:131495. DOI: 10.1016/j.colsurfa.2023.131495

[42] Dewen S, Liqun L, Xiaodong Z, Zhiming S, Qianping R, Pingping H, et al. Fluorine-Modified Hyperbranched Waterborne Polyurethane Resin and Waterborne Nano Anti-Corrosion Self-Cleaning Finish Paint Using Same. China: Jiangsu Sobute New Mat Co Ltd; Jiangsu Fengcai Building Mat Group Co Ltd.; 2021 CN113683748A

[43] Wu Y, Shen Y, Tao J, He Z, Xie Y, Chen H, et al. Facile spraying fabrication of highly flexible and mechanically robust superhydrophobic F-SiO_2@PDMS coatings for self-cleaning and drag-reduction applications. New Journal of Chemistry. 2018;**42**(22):18208-18216. DOI: 10.1039/C8NJ04275F

[44] Chen L, Chen S, Wang L, Yang S. Super-Hydrophobic Polyurethane/Oxide Nano Particle Hybrid Coating Material and Preparation Method Thereof. China: Univ Nanjing; 2010 CN101838496A

[45] Zhu GJ, Ren PG, Wang J, Duan Q, Ren F, Xia WM, et al. A highly sensitive and broad-range pressure sensor based on polyurethane mesodome arrays embedded with silver nanowires. ACS Applied Materials & Interfaces. 2020;**12**(17):19988-19999. DOI: 10.1021/acsami.0c03697

[46] Cherupurakal N, Mozumder MS, Mourad AHI, Lalwani S. Recent advances in superhydrophobic polymers for antireflective self-cleaning solar panels. Renewable and Sustainable Energy Reviews. 2021;**151**:111538. DOI: 10.1016/j.rser.2021.111538

[47] Chen K, Zhou J, Ge F, Zhao R, Wang C. Smart UV-curable fabric coatings with self-healing ability for durable self-cleaning and intelligent oil/water separation. Colloids and Surfaces A: Physicochemical and Engineering Aspects. 2019;**565**:86-96. DOI: 10.1016/j.colsurfa.2019.01.003

Chapter 4

A Self-Cleaning Approach Utilizing Metal Oxide Thin Films and Nanocomposites

Abdullah Al Nahid

Abstract

Self-cleaning technology mimics the natural self-cleaning abilities of plants and animals such as lotus effect, to create a surface that is hydrophobic and oleophobic, meaning it repels water and oil. The resultant surface is resistant to dirt and grime, making it easier to clean and maintain, reducing labor costs and time consumption. However, it is not only limited to the superhydrophobic surface for making the water roll off instead of sliding but also modern research focuses on incorporating photocatalysts to break down organic compounds during daylight at outdoor applications. In addition, self-cleaning surfaces and coatings are attracting research attention due to their ability to self-disinfect. This review highlights the use of metal oxide-based nanocomposite for self-cleaning purposes. This chapter provides an outlook of different metal oxide and metal-metal oxide nanocomposites in advancing self-cleaning properties, durability, and other mechanical properties. This chapter aims to give a general overview of a variety of polymeric metal oxide-based systems and methods that enhance self-cleaning behavior as well as the projection toward future research.

Keywords: hydrophobicity, metal oxide, nanocomposite, self-cleaning, photocatalytic activity

1. Introduction

Self-cleaning materials have become increasingly popular in recent years due to their remarkable ability to remove dirt, contaminants, and even microorganisms from their surfaces. The materials hold great promise for a variety of applications, including architectural coatings, automotive surfaces, medical devices as well as self-cleaning exterior construction materials, interior furnishing materials, road construction materials, solar cells, car mirrors, textiles, utensils, roof tiles, etc. Among the various strategies employed to achieve self-cleaning properties, the use of metal oxide thin films and nanocomposites has emerged as a particularly promising approach.

Metal oxides, such as titanium dioxide (TiO_2), zinc oxide (ZnO), and iron oxide (Fe_2O_3), have been extensively studied for their photocatalytic properties and their ability to generate reactive oxygen species (ROS) upon exposure to ultraviolet (UV) light. These ROS possess strong oxidizing power and can break down organic compounds, rendering them highly effective in the degradation of dirt, organic pollutants, and even bacteria and viruses [1]. For example, ZnO is an n-type semiconductor with band gap E_g = 3.37 eV. CuO is a p-type, narrow band gap (E_g = 1.2 eV) semiconductor capable of large absorption of light in spite of its low photocatalytic efficiency [2]. Through their light absorption capabilities, incorporating metal oxide thin films onto surfaces or nanoparticles into composite materials can impart self-cleaning properties to a wide range of substrates.

In this book chapter, we delve into the world of self-cleaning materials and explore the potential of metal oxide thin films and nanocomposites for achieving this functionality. We begin by providing a comprehensive overview of the principles underlying self-cleaning mechanisms and the unique properties of metal oxides that make them ideal candidates for self-cleaning applications. We discuss the photocatalytic activity of metal oxides, their role in generating ROS, and the mechanisms through which dirt and contaminants are degraded and removed from the surface.

Moreover, we explore the diverse range of applications that can benefit from self-cleaning materials. From self-cleaning glass windows that maintain their transparency and clarity to self-cleaning textiles that repel stains and odors, the potential of metal oxide thin films and nanocomposites is vast. We discuss the challenges and opportunities associated with scaling up these materials for real-world applications and address the environmental implications of self-cleaning technologies.

This review aims to provide an in-depth understanding of the self-cleaning approach utilizing metal oxide thin films and nanocomposites. By exploring the fundamental principles, synthesis techniques, and potential applications, we hope to contribute to the advancement of self-cleaning materials and inspire further research and development in this exciting field.

2. Origin of self-cleaning technology

The history of nature-inspired self-cleaning can be traced back to the early twentieth century when scientists and engineers began studying natural phenomena that exhibited self-cleaning properties. Self-cleaning refers to the spontaneous removal of contaminants, such as dirt, organic compounds, and pollutants, from the surface of a material without the need for external cleaning agents or mechanisms. However, in the 1960s, Swiss engineer Georges de Mestral invented Velcro, a fastening system inspired by the way burrs clung to his clothes and dog's fur. This innovation sparked interest in biomimicry, the practice of emulating nature's designs and processes in engineering and design [3].

In the 1970s, Lotus leaf's self-cleaning ability became a subject of scientific inquiry. Researchers discovered that the leaf's microscale surface structure, consisting of tiny wax crystals, repelled water and prevented dirt particles from adhering [4]. This discovery laid the foundation for the development of superhydrophobic and self-cleaning surfaces.

3. Mechanism of self-cleaning

There are two mechanisms responsible for surface cleaning, that is, hydrophilic and hydrophobic.

3.1 Hydrophilic and superhydrophilic surfaces

Hydrophilic and superhydrophilic surfaces utilize the photocatalytic activity of the coated materials in achieving the self-cleaning property. A surface is said to be hydrophilic if water spread over the surface evenly without the formation of droplets. Usually ordinary surfaces with typical wetting have contact angle 30° to 90°, whereas surfaces with contact angles less than 30° are termed hydrophilic, since the forces of interaction between water and the surface are almost equal to the forces of bulk water and water does not drain cleanly from these surfaces. However, when the advancing contact angle of water on the surface is less than 5°–10° [5–7] or the time for complete moisture to be achieved by small drops of water has to be under 0.5 seconds then the surface is designated as superhydrophilic [8]. However, superhydrophilic surfaces have a strong affinity for water due to the low surface energy and the micro/nanostructured surface of the metal oxide thin film [9]. As the water spreads over a surface, dirt particles are picked up and carried away, effectively cleaning the area.

Nevertheless, photocatalytic self-cleaning metal oxide thin films, such as titanium dioxide (TiO_2) and zinc oxide (ZnO), exhibit photocatalytic properties, enabling them to undergo self-cleaning under light. When exposed to ultraviolet (UV) or visible light, the metal oxide thin film absorbs photons and generates electron-hole pairs. The holes at the valence band (VB) can accept electrons from the surrounding moisture and generate hydroxyl radicals ($OH^•$) (**Figure 1c**) which are capable of acting as a detergent for the troposphere and breaking down organic pollutants into small fragments with low adhesion to the surface. Furthermore, excited electrons in the conduction band (CB) can react with atmospheric oxygen and produce superoxide radicals ($O_2^{•-}$) (**Figure 1b**) that can oxidize organic compounds and break them down into harmless compounds. This photocatalytic oxidation process effectively removes contaminants from the surface but generates highly reactive oxygen species (ROS), such as hydroxyl radicals ($OH^•$) and superoxide radicals ($O_2^{•-}$) resulting in self-cleaning. As a result, either a sprinkle of water or rainwater can easily remove the dirt from the surface without using any laborious cleaning.

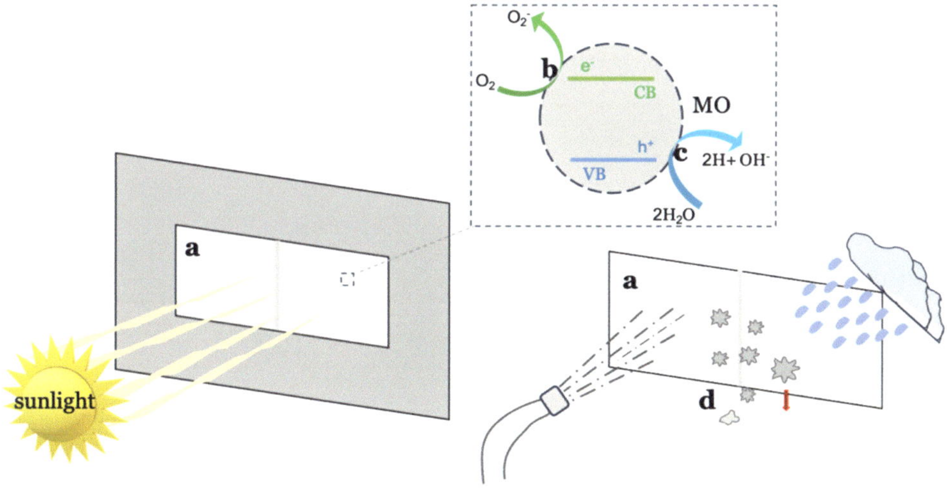

Figure 1.
Illustration of photocatalytic self-cleaning mechanism of MO thin film. (a). Window glass coated with MO thin film; (d). using rainwater or sprinklers can remove dirt with no effort or extensive cleaning.

Furthermore, the thin film promotes self-drying through the photocatalytic generation of hydroxyl radicals, which dissolve water molecules, further, eliminate contaminants, and facilitate quick evaporation. The photocatalytic self-cleaning action of metal oxide thin film can be simply illustrated as shown in **Figure 1**.

3.2 Superhydrophobic self-cleaning

Another approach to achieving self-cleaning property is to create a surface superhydrophobic and a water contact angle above 150° with a sliding age below 5° [10, 11]. Normal hydrophobic surfaces have a water contact angle between 90° to 120°. However, in order to make a surface super hydrophobic, coating and nanostructure engineering are required. For this purpose, polymer nanocomposites can be designed to exhibit this property by incorporating hydrophobic nanoparticles or modifying the surface with a hydrophobic coating. A superhydrophobic self-cleaning mechanism is inspired by nature, specifically the lotus leaf [4] and certain insect wings [12], which are extremely water-repellent. These mechanisms are designed to repel water and prevent the adhesion of dirt and other contaminants, enabling surfaces to remain clean with minimal maintenance [13].

When water droplets come into contact with a superhydrophobic surface, they form spherical beads and roll off, carrying away dirt and contaminants with them. As water droplets roll off the superhydrophobic surface, they pick up and carry away any dust, dirt, or contaminants present on the surface. The low adhesion of the water droplets to the surface prevents the contaminants from sticking, effectively cleaning the surface in the process. **Figure 2** shows a visual representation of this phenomenon.

Superhydrophobic coatings can be applied to many everyday surfaces, such as windshields, bodies of vehicles, windows and doors, skyscrapers, solar cell panels, fabrics, sports shoes, metals, paper, sponges, woods, marbles, and the list is endless [14]. It will be advantageous for the architecture, automotive, electronics, and textile industries to use superhydrophobic coatings in the future, as they will increase their profitability, lucrativeness, and durability by preventing fouling due to contaminants.

The surface to be made superhydrophobic is coated with a material that has low surface energy, such as a hydrophobic coating or a nanoparticle-based solution. This coating alters the surface properties, making it highly water-repellent. Then the

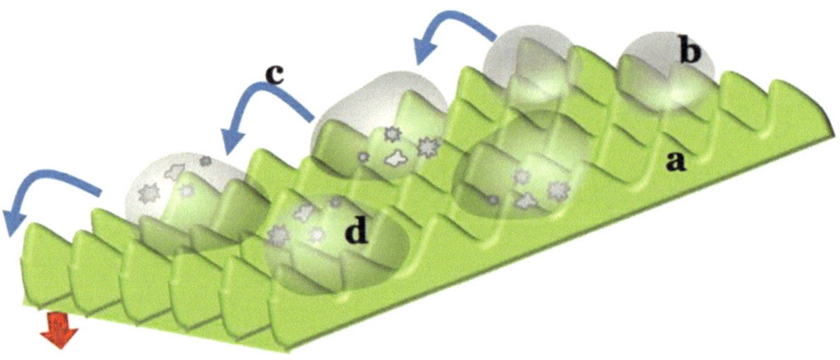

Figure 2.
Visual representation of self-cleaning action of a superhydrophobic surface. (a). Rough surface resembling a lotus leaf; (b). water droplets. (c & d). Rolling off water droplets while capturing dirt inside and draining away.

surface is engineered with microscale or nanoscale structures, such as micro-ridges or nanopillars [15]. These structures create a rough surface texture, increasing the effective surface area and reducing contact between water droplets and the surface. The combination of the surface coating and the micro/nanostructures mimics the lotus leaf's characteristics [3]. When water droplets come into contact with the super-hydrophobic surface, they tend to bead up and roll off due to the reduced contact area and the low surface energy of the coating. This behavior is known as the "lotus effect."

4. Examples of some metal oxides self-cleaning

The self-cleaning mechanism of TiO_2 can be applied in various applications, such as building materials that includes paints, coatings, or materials to create self-cleaning surface. TiO_2 coatings can be applied to solar panels to prevent the accumulation of dust and dirt, which can decrease the efficiency of the panels. The self-cleaning property of TiO_2 ensures that the panels remain clean and maximize their energy generation [16]. TiO_2 coatings on glass surfaces make them self-cleaning. This application is particularly useful for windows, facades, and glass surfaces in outdoor environments. TiO_2 nanoparticles can be used in air filters or coatings to remove airborne pollutants and improve indoor air quality. It is worth noting that the self-cleaning efficiency of TiO_2 can vary depending on factors such as the specific TiO_2 formulation, the intensity of UV light, the presence of contaminants, and the surface area of TiO2 exposed to the light.

Other semiconductor materials, such as zinc oxide (ZnO), tungsten oxide (WO_3), and cadmium selenide (CdSe), are all examples of self-cleaning [17]. Different classes of metal oxides used for self-cleaning coating material can be categorized as follows, as shown in **Figure 3**.

The photocatalytic property of most metal oxides can be greatly enhanced by combining them with other metal oxides, even though few metal oxides do not possess self-cleaning properties. A good example is ZrO_2, which is not intrinsically self-cleaning but can be used as part of composite systems or modified to act as catalysts [18]. Furthermore, metal-doped metal oxides such as aluminum doped zinc oxide (AZO) and copper oxide (CuO) thin film may have potential applications in photovoltaics like solar cells [19, 20]. Other metal ions such as Cr, V, Fe, Pb, Cu, and Al [17] and nonmetals like C, N, and F doped metal oxide like TiO_2 have shown potential self-cleaning surfaces [21]. The self-cleaning ability helps to maintain the catalytic activity by continuously removing adsorbed contaminants, resulting in efficient and long-lasting remediation processes. Therefore, metal oxide thin films may have potential applications in environmental remediation and solar panels. These panels can maintain optimal light absorption and maximize their energy by incorporating self-cleaning metal oxide thin films.

Nevertheless, the self-cleaning properties of metal oxides such as $ZrO_2 + TiO_2$ and $Al_2O_3 + ZrO_2$ nanocomposites have found numerous applications in various fields [18, 20]. In architecture, self-cleaning coatings on building facades can maintain their esthetic appearance and reduce maintenance costs. In automotive and aerospace industries, self-cleaning coatings protect surfaces from environmental contaminants, improving performance and longevity. Additionally, self-cleaning metal oxide nanocomposites have shown promise in water purification, air pollution control, and biomedical applications.

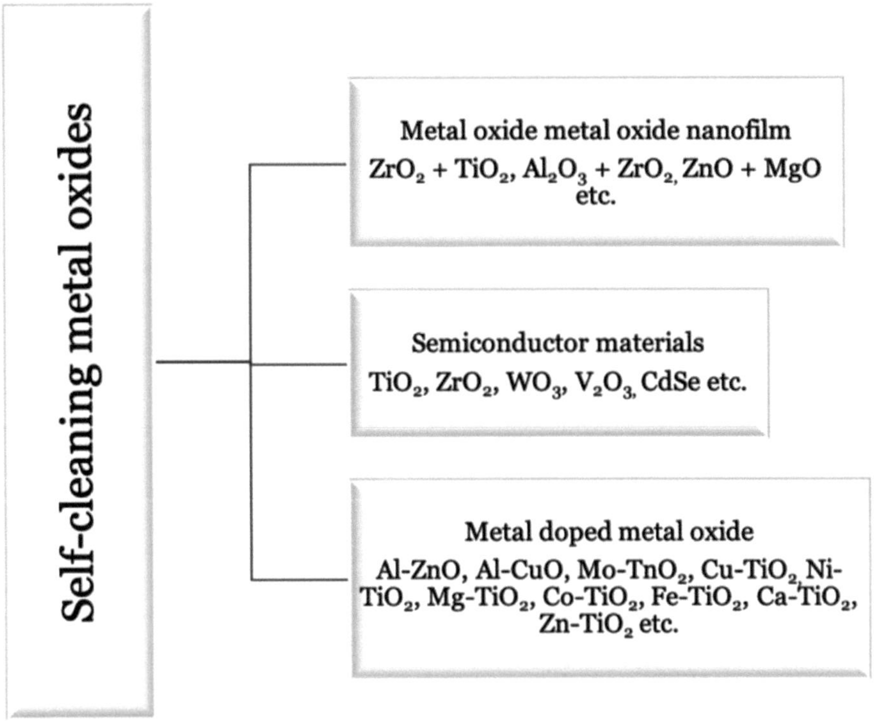

Figure 3.
Various types of metal oxides for self-cleaning coating.

5. Examples of some polymer nanocomposites self-cleaning

Polymer nanocomposites, which consist of polymer matrices reinforced with nanoparticles, offer a promising solution for the development of self-cleaning materials. This chapter explores some of the important polymer nanocomposites used for self-cleaning purposes, their properties, fabrication methods, and potential applications. Polymer nanocomposites provide an avenue to replicate the nature inspire self-cleaning properties on a commercial scale. However, important polymer nanocomposites can be categorized as shown in **Figure 4**, and their details are discussed in Sections 5.1–5.4 focusing on the self-cleaning properties.

5.1 Silica-based polymer nanocomposites

Silica nanoparticles are widely employed in polymer nanocomposites for their hydrophobic and oleophobic properties. These nanoparticles are incorporated into polymer matrices such as polyethylene, polypropylene, or poly (methyl methacrylate) (PMMA), hyperbranched polyester acrylate to enhance their surface energy and reduce adhesion of contaminants. Silica-based nanocomposites exhibit excellent self-cleaning properties due to their low surface energy and high contact angle with water and oils [22].

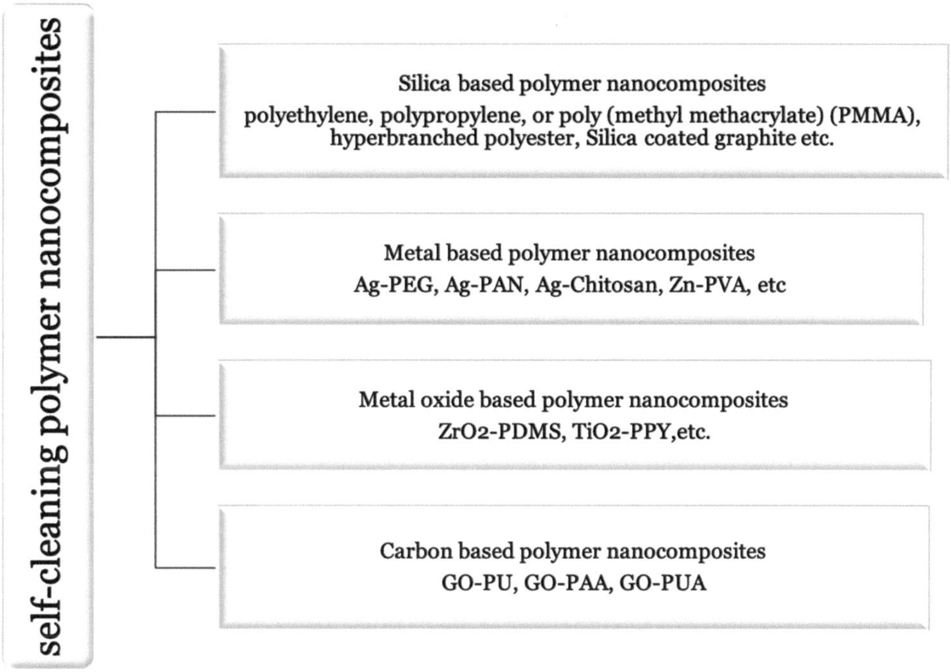

Figure 4.
Various polymer nanocomposites for self-cleaning coating.

5.2 Metal-based polymer nanocomposites

Metal-based polymer nanocomposites are widely used for self-cleaning due to their antibacterial and catalytic activity. Silver-polyethylene glycol (Ag-PEG) exhibits antimicrobial properties, Ag-chitosan shows antifungal properties, Ag-polyacrylonitrile (Ag-PAN) offers antibacterial and antimicrobial activity, ultraviolet blocking, and catalytic activities [23, 24]. Zinc can also be combined with chitosan, cellulose, and polyvinyl alcohol and is shown to have photocatalytic and antimicrobial properties, making it widely used in food contact surfaces and packaging [25]. The combination of gold with several polymers such as chitosan, polyoxoborate matrix, heparin-polyvinyl alcohol, and carboxyl methylcellulose exhibits antimicrobial and wound healing properties that can be used in biomedicine, medical devices, and sensors as well as photovoltaics [26, 27].

5.3 Metal oxide-based polymer nanocomposites

Metal oxide-based polymer nanocomposites such as A coating made of Aramid zyrconia nanocomposites, that is, ZrO_2 combined with 1, 4-phenylene diamine, demonstrated remarkable protection by controlling charge transfer at the interface between steel alloy and electrolyte, preventing alloy dissolution [28]. TiO_2 and polyurethane nanocomposites have shown self-cleaning and antibacterial activities of textiles [29, 30]. Moreover, polydimethylsiloxane (PDMS) incorporated with ZrO_2 can produce superhydrophobic surfaces on cotton fabrics [31]. Additionally, the photocatalytic printed film of titania polypyrrole (PPy/TiO_2) nanocomposites shows excellent photocatalytic, self-cleaning, and antibacterial functionalities [32]. It is, therefore, possible to create self-cleaning surfaces using a suitable polymer together with other metal oxides.

5.4 Carbon-based polymer nanocomposites

Carbon-based nanomaterials such as graphene, carbon nanotubes (CNTs), and their derivatives, nanodiamond and fullerenes, offer unique self-cleaning properties due to their low surface energy, high electrical conductivity, and high mechanical strength. Carbonaceous materials have been used in polymer matrixes as filler or reinforcement to form carbon polymer composites. When incorporated into polymer matrices, carbonaceous nanomaterials can create self-cleaning surfaces that repel water, oils, and dirt particles. Additionally, their conductivity allows for generating electrostatic charges, which can repel contaminants from the surface. However, several carbon-based polymer nanocomposites include reduced graphene oxide-polyurethane (GO-PU), graphene oxide-polyacrylic acid (GO-PAA), graphene oxide-polyurethan acrylate (GO-PUA) coating, etc. [33].

Polymer nanocomposites derived from carbon offer the potential for individual and hybrid development of a wide range of innovative materials for a wide range of applications, such as thermal materials, shielding for electromagnetic interfaces, sensors, and energy storage [34].

However, polymer nanocomposite coating has some advantages because polymers are excellent host materials for nanoparticles. When the nanoparticles are embedded or encapsulated in polymer, the polymer acts as a surface capping agent and generates proper coating adhesion to the surface [35]. Metal/metal oxide nanoparticle-polymer composites are versatile materials containing dispersion of nanosized filler materials within the polymer matrix. Owing to their nanoscale size features and high surface-to-volume ratios, they possess unique combination of multifunctional properties such as barrier properties and high ductility without strength loss. For example, if a drop of water trying to get through the film made with nanocomposite, it would face more barriers than in case of conventional composites because the distance between fillers in nanocomposite is much smaller. Additionally, Embedding nanoparticles inside polymer thin films facilitates immobilization and organization of the metal nanoparticles and tuning of their electronic and optical responses by the dielectric environment.

6. Overview of preparing nanocomposites and its fabrications

The most convenient routes of embedding the metallic nanoparticles into the polymer involve the following two ways. In situ method – growth of nanoparticles into polymer matrices through reduction or decomposition of precursors; ex situ method – incorporation of premade particles into a premade polymer matrix using a common blending solvent. In situ polymerization involves the simultaneous synthesis of polymer and nanoparticle formation within a reaction medium. This method allows for better control over nanoparticle dispersion and interfacial interactions, leading to enhanced self-cleaning properties. In situ polymerization is particularly suitable for the synthesis of polymer nanocomposites with specialized applications.

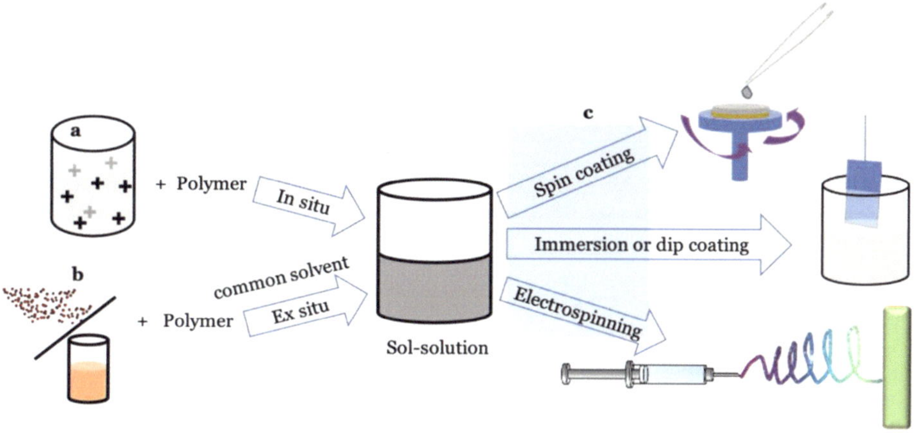

Figure 5.
Graphical demonstration of the general routes of polymer nanocomposite preparation and its popular fabrication processes. (a). Nanomaterials precursor solution; (b). prepared nanoparticles or nano-colloid; (c). few techniques for nanocomposite fabrication.

Whereas in ex situ method, solution mixing involves dispersing nanoparticles in a solvent and subsequently adding the polymer matrix to form a homogeneous mixture. The mixture is then cast or coated onto a substrate and dried to form a nanocomposite film. Solution mixing is a simple and scalable method for producing polymer nanocomposites with self-cleaning properties. Either method can be suitable for preparing composite material depending on the criteria of polymer and nanomaterials. There are also several easy and inexpensive ways to apply these coatings including sol–gel, immersion, spin coating, printing, and others. A general overview of nanocomposites preparation and its fabrication can be outlined as illustrated in **Figure 5**.

7. Challenges and future directions

The main challenges were twofold for engineering nanostructures in order to get better properties. First, the addition of nanosized particles into low-viscosity oligomers usually leads to a problematic liquid-to-solid transition at very low particle loadings, which may have a catastrophic effect on the low-pressure, nanoscale replication fidelity. Secondly, the dense network of nanoparticles may act as a filter for the migration of the surfactant molecules and therefore suppress the superhydrophobic effect.

It is true that metal oxide thin films and polymer nanocomposite materials possess unique properties that enable them to repel dirt, water, and other contaminants, resulting in less maintenance and cleaning. For their self-cleaning capabilities to be further enhanced, however, there are several challenges to overcome, as described below, and **Figure 6** depicts its summary outline.

7.1 Durability and stability

One of the primary challenges is ensuring the long-term durability and stability of metal oxide thin films and polymer nanocomposites. These materials may undergo degradation over time due to exposure to environmental factors, such as UV radiation,

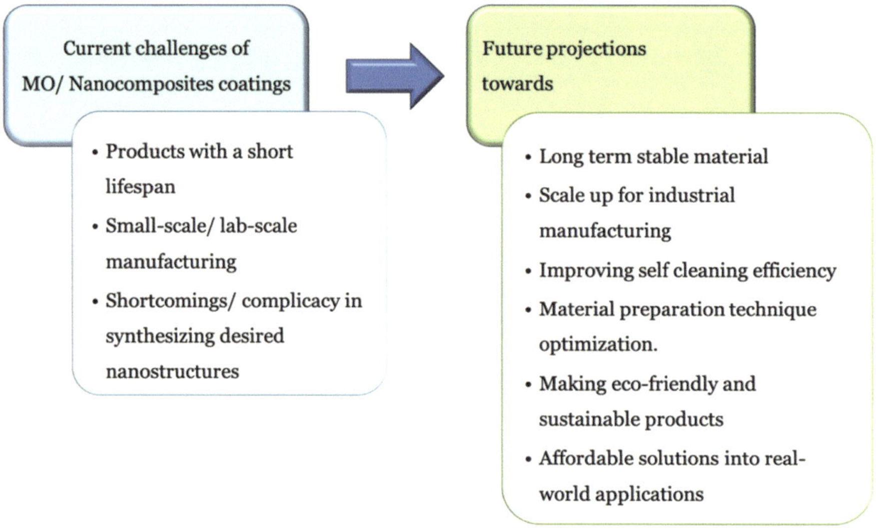

Figure 6.
A simple overview of current challenges and future projections for enhancing properties of MO/nanocomposites.

humidity, and temperature fluctuations. Future research should focus on improving the stability of these materials to maintain their self-cleaning properties over extended periods.

7.2 Scalability and manufacturing

To make self-cleaning materials commercially viable, it is crucial to develop scalable manufacturing processes. Currently, the fabrication techniques for metal oxide thin films and polymer nanocomposites often involve complex and expensive procedures, limiting their large-scale production. Future directions should explore cost-effective and scalable manufacturing methods that can be easily implemented in various industries.

7.3 Enhanced self-cleaning efficiency

While metal oxide thin films and polymer nanocomposites demonstrate self-cleaning properties, there is room for improvement in terms of efficiency. Researchers can explore novel nanostructures, surface modifications, or coating techniques to optimize the self-cleaning effect. Increasing the contact angle, reducing surface energy, and enhancing the photocatalytic activity of metal oxide films are some potential avenues to pursue.

7.4 Selective self-cleaning

Current self-cleaning materials typically repel both water and oil-based contaminants. However, there may be applications where selective self-cleaning is desired. Future research could focus on developing materials with tailored properties to selectively repel specific types of contaminants, allowing for targeted cleaning capabilities in specific environments or industries.

7.5 Environmental considerations

As with any emerging technology, it is crucial to consider the environmental impact of self-cleaning materials. Future research should focus on developing sustainable and eco-friendly fabrication processes, reducing the use of harmful chemicals, and exploring recyclability and biodegradability of these materials.

7.6 Real-world applications

While self-cleaning materials have been studied extensively in the lab, their practical implementation in real-world applications is still limited. Future research should aim to bridge this gap by focusing on developing prototypes and conducting field tests in various industries, such as architecture, automotive, and healthcare, to demonstrate the feasibility and benefits of using self-cleaning materials in different settings.

8. Conclusion

Metal oxide thin film and polymer nanocomposites hold great promise as self-cleaning materials, offering a range of applications and potential benefits due to having several advantages including flexibility, lightweight, and ease of processing. However, there are still challenges that need to be addressed for widespread practical implementation. Continued research and development in this field such as enhancing the efficiency of photocatalytic processes, optimizing stability, and exploring sustainable synthesis, are necessary to overcome existing challenges and unlock the full potential of these materials for practical use in various industries. Continued research and innovation will drive the development of advanced self-cleaning materials with enhanced efficiency and broader practicality in the near future.

Conflict of interest

The authors declare no conflict of interest.

Author details

Abdullah Al Nahid
Bangladesh Council of Science and Industrial Research, Dhaka, Bangladesh

*Address all correspondence to: aan1985du@gmail.com

References

[1] Dogra V, Verma D, Kishore C. A Prospective Utilization of Metal Oxides for Self-Cleaning and Antireflective Coatings. USA: Elsevier Inc.; 2022. pp. 139-152. DOI: 10.1016/B978-0-323-90410-0.00008-8

[2] Upadhaya D, Dhar PD. Self-cleaning activity of CuO/ZnO Heterostructure: A synergy of Photocatalysis and hydrophilicity. Journal of the Taiwan Institute of Chemical Engineers. 2022;**132**:104216. DOI: 10.1016/j.jtice.2022.104216

[3] Collins CM, Safiuddin M. Lotus-leaf-inspired biomimetic coatings: Different types, key properties, and applications in infrastructures. Infrastructures. 2022;**7**(4):46. DOI: 10.3390/infrastructures7040046

[4] Latthe SS, Terashima C, Nakata K, Fujishima A. Superhydrophobic surfaces developed by mimicking hierarchical surface morphology of lotus leaf. Molecules. 2014;**19**(4):4256-4283. DOI: 10.3390/molecules19044256

[5] Samanta A, Wang Q, Shaw SK, Ding H. Roles of chemistry modification for laser textured metal alloys to achieve extreme surface wetting behaviors. Materials and Design. 2020;**192**:108744. DOI: 10.1016/j.matdes.2020.108744

[6] Drelich J, Chibowski E. Superhydrophilic and Superwetting surfaces: Definition and mechanisms of control. Langmuir. 2010;**26**(24):18621-18623. DOI: 10.1021/la1039893

[7] Drelich J, Chibowski E, Meng DD, Terpilowski K. Hydrophilic and superhydrophilic surfaces and materials. Soft Matter. 2011;**7**(21):9804-9828. DOI: 10.1039/c1sm05849e

[8] Law K-Y. Definitions for hydrophilicity, hydrophobicity, and Superhydrophobicity: Getting the basics right. Journal of Physical Chemistry Letters. 2014;**5**(4):686-688. DOI: 10.1021/jz402762h

[9] Nundy S, Ghosh A, Mallick TK. Hydrophilic and Superhydrophilic self-cleaning coatings by morphologically varying ZnO microstructures for photovoltaic and glazing applications. ACS. Omega. 2020;**5**(2):1033-1039. DOI: 10.1021/acsomega.9b02758

[10] Zhang Y, Zhang Z, Yang J, Yue Y, Zhang H. A review of recent advances in Superhydrophobic surfaces and their applications in drag reduction and heat transfer. Nanomaterials. 2022;**12**(1):44. DOI: 10.3390/nano12010044

[11] Starinskiy S, Starinskaya E, Miskiv N, Rodionov A, Ronshin F, Safonov A, et al. Spreading of impacting water droplet on surface with fixed microstructure and different wetting from Superhydrophilicity to Superhydrophobicity. Water (Switzerland). 2023;**15**(4):719. DOI: 10.3390/w15040719

[12] Sanjay SL, Annaso BG, Chavan SM, Rajiv SV. Recent Progress in preparation of Superhydrophobic surfaces: A review. Journal of Surface Engineered Materials and Advanced Technology. 2012;**2**:76-94. DOI: 10.4236/jsemat.2012.22014

[13] Ge-Zhang S, Cai T, Yang H, Ding Y, Song M. Biology and nature: Bionic Superhydrophobic surface and principle. Frontiers in Bioengineering and Biotechnology. 2022;**10**:1-19. DOI: 10.3389/fbioe.2022.1033514

[14] Latthe SS, Sutar RS, Kodag VS, Bhosale AK, Kumar KAM, Sadasivuni K,

et al. Self – Cleaning Superhydrophobic coatings: Potential industrial applications. Progress in Organic Coatings. 2019;**128**:52-58. DOI: 10.1016/j.porgcoat.2018.12.008

[15] Wang Y, Zhang Q, Li P, Huang JT. A durable and sustainable Superhydrophobic surface with intertwined cellulose/SiO2 blends for anti-icing and self-cleaning applications. Materials and Design. 2022;**217**:110628. DOI: 10.1016/j.matdes.2022.110628

[16] Isaifan RJ, Samara A, Suwaileh W, Johnson D, Yiming W, Abdallah AA, et al. Improved self-cleaning properties of an efficient and easy to scale up TiO2 thin films prepared by adsorptive self-assembly. Scientific Reports. 2017;**7**(1):1-9. DOI: 10.1038/s41598-017-07826-0

[17] Topçu IB. Self-cleaning concretes: An overview. Journal of Cement Based Composites. 2020;**1**(2):6-12. DOI: 10.36937/cebacom.2020.002.002

[18] Rudakova AV, Emeline AV, Bulanin KM, Chistyakova LV, Maevskaya MV, Bahnemann DW. Self-cleaning properties of zirconium dioxide thin films. Journal of Photochemistry and Photobiology A: Chemistry. 2018;**367**:397-405. DOI: 10.1016/j.jphotochem.2018.08.037

[19] Stella A, Nisoli M, de Silvestri S, Svelto O, Lanzani G, Cheyssac P, et al. Size effects in the ultrafast electronic dynamics of metallic tin nanoparticles. Physical Review B. 1996;**53**:15497

[20] Nordseth Ø, Kumar R, Bergum K, Fara L, Dumitru C, Craciunescu D, et al. Metal oxide thin-film heterojunctions for photovoltaic applications. Materials (Basel). 2018;**11**(12):1-10. DOI: 10.3390/ma11122593

[21] Krishna MG, Vinjanampati M, Purkayastha DD. Metal oxide thin films and nanostructures for self-cleaning applications: Current status and future prospects. EPJ. Applied Physics. 2013;**62**(3):300011-300012. DOI: 10.1051/epjap/2013130048

[22] Cully P, Karasu F, Müller L, Jauzein T, Leterrier Y. Self-cleaning and Wear-resistant polymer nanocomposite surfaces. Surface and Coatings Technology. 2018;**348**:111-120. DOI: 10.1016/j.surfcoat.2018.05.040

[23] Zarnegar Z, Safari J, Zahraei Z. Design, synthesis and antimicrobial evaluation of silver decorated magnetic polymeric nanocomposites. Nano-Structures & Nano-Objects. 2019;**19**:100368. DOI: 10.1016/j.nanoso.2019.100368

[24] Rehan M, Nada AA, Khattab TA, Abdelwahed NAM, El-Kheir AAA. Development of multifunctional Polyacrylonitrile/silver nanocomposite films: Antimicrobial activity, catalytic activity, electrical conductivity, UV protection and SERS-active sensor. Journal of Materials Research and Technology. 2020;**9**(4):9380-9394. DOI: 10.1016/j.jmrt.2020.05.079

[25] Wen Y-H, Tsou C-H, de Guzman MR, Huang D, Yu Y-Q, Gao C, et al. Antibacterial nanocomposite films of poly(vinyl alcohol) modified with zinc oxide-doped multiwalled carbon nanotubes as food packaging. Polymer Bulletin. 2022;**79**(6):3847-3866. DOI: 10.1007/s00289-021-03666-1

[26] Adil SF, Shaik MR, Nasr FA, Alqahtani AS, Ahmed MZ, Qamar W, et al. Enhanced apoptosis by functionalized highly reduced graphene oxide and gold nanocomposites in MCF-7 breast cancer cells. ACS. Omega. 2021;**6**(23):15147-15155. DOI: 10.1021/acsomega.1c01377

[27] Zhang K, Zhao G. An effective wound healing material based on gold

incorporation into a heparin-polyvinyl alcohol nanocomposite: Enhanced In vitro and In vivo Care of Perioperative Period. Journal of Cluster Science. 2022;**33**(4):1655-1665. DOI: 10.1007/s10876-021-02078-5

[28] Nazeer AA, Al Sagheer F, Bumajdad A. Aramid-zirconia nanocomposite coating with excellent corrosion protection of stainless steel in saline media. Frontiers in Chemistry. 2020;**8**:1-11. DOI: 10.3389/fchem.2020.00391

[29] Soytaş SH, Oğuz O, Menceloğlu YZ. Polymer Nanocomposites With Decorated Metal Oxides. In: Polymer Composites with Functionalized Nanoparticles Synthesis, Properties, and Applications. USA: Elsevier Inc.; 2018. DOI: 10.1016/B978-0-12-814064-2.00009-3

[30] Bukit BF, Frida E, Humaidi S, Sinuhaji P. Selfcleaning and antibacterial activities of textiles using nanocomposite oil palm boiler ash (OPBA), TiO2 and chitosan as coating. South African. Journal of Chemical Engineering. 2022;**41**:105-110. DOI: 10.1016/j.sajce.2022.05.007

[31] Eduok U. New Superhydrophobic and self-cleaning zirconia/polydimethylsiloxane nanocomposite coated cotton fabrics. New Journal of Chemistry. 2021;**45**(2):638-650. DOI: 10.1039/d0nj04848h

[32] Nejad SAT, Soleimani-Gorgani A, Pishvaei M. Multifunctional screen-printed films using polymer nanocomposite based on PPy/TiO2: Conductive, photocatalytic, self-cleaning and antibacterial functionalities. Iranian Polymer Journal (English Ed.). 2023;**32**(6):647-659. DOI: 10.1007/s13726-023-01153-0

[33] Siwal SS, Zhang Q, Devi N, Thakur VK. Carbon-based polymer nanocomposites for high performance energy storage application. Polymers. 2020;**12**(3):1-31. DOI: 10.3390/polym12030505

[34] Paul DR, Robeson LM. Polymer nanotechnology: Nanocomposites. Polymer (Guildf). 2008;**49**(15):3187-3204. DOI: 10.1016/j.polymer.2008.04.017

[35] Caseri WR. In situ synthesis of polymer-embedded nanostructures. In: Nanocomposites: In Situ Synthesis of Polymer-Embedded Nanostructures. USA: John Wiley & Sons, Inc.; 2013. pp. 45-72. DOI: 10.1002/9781118742655.ch2

Chapter 5

Plasma Based Approaches to Achieve Self-Cleaning Surfaces

Deepanjana Adak and Raghunath Bhattacharyya

Abstract

The role of gaseous plasma has proven to be very beneficial in creating self-cleaning of various surfaces. Few references are there, in the published literature, on plasma enhanced hydrophilicity/hydrophobicity behavior of surfaces. A range of atmospheric pressure plasma spray systems are gaining popularity for creating self-cleaning surfaces, with some unique features, as also to fabricate new types of self-cleaning materials. In this chapter a brief introduction to essentials of plasma processing will be first presented, followed by examples of plasma assisted surface modification. This will include plasma cleaning, plasma etching, plasma polymerization/deposition, etc. Subsequently, various plasma assisted techniques to achieve a variety of self-cleaning surfaces will be highlighted. A unique combination of plasma-based approaches and sol–gel derived coating will also be discussed.

Keywords: self-cleaning, atmospheric-plasma, cold-plasma, superhydrophilic, superhydrophobic

1. Introduction

1.1 What is plasma?

> *The term has been first used to describe ionized gas by Irving Langmuir in the year 1929. Plasma can be defined as a quasi-neutral gas of charged and neutral particles characterized by collective behavior.*

Depending on the energy of particle constituting matter, the matter can be broadly divided into 3 states – solid, liquid, and gases. However there exist energetically fourth state of matter apart from these three states, known as, *Plasma*. Generally, the transformation from one state to another is achieved in the presence of energy sufficient enough to overcome intermolecular forces. Likewise the transformation to fourth state of matter, plasma state, from gaseous state takes place when the gas molecules are heated to extremely high temperature or subject to high energy radiation [1]. The most common examples of species that exist in plasma state are interstellar matter such as sun and other stars. The primary benefit of plasma-assisted solid processing arises from the ability to conduct chemical reactions at substrate temperatures significantly lower than those achievable through standard reactions occurring at thermal equilibrium [2].

Plasma, with its fascinating characteristics, arises from the electromagnetic interactions among charged particles. It comprises electrons, ions, neutrals, electromagnetic radiation (photons), and electromagnetic fields, all contributing to its unique properties [3–9]. One notable attribute of plasma is its striking quasi-neutrality, where the positive and negative charges are intricately balanced, leading to an overall state of electrical neutrality [10].

Plasma can be broadly classified into two groups based on temperature and equilibrium conditions. The first group is thermal or equilibrium plasma, characterized by high ionization and thermal equilibrium among charged particles at temperatures around 10^4 K. The second group is non-thermal or non-equilibrium plasma, commonly known as cold plasma. In cold plasma, ionization is milder, and electrons do not reach thermal equilibrium with heavier particles. As a result, the gas temperature remains close to room temperature or slightly elevated, typically ranging from 40 to 150°C. Additionally, the electron density in cold plasma is typically below 1000 m^{-3} [11]. Non-thermal or cold plasmas can be generated under both low-pressure and atmospheric-pressure conditions. Generally, the temperature of each species within the plasma is determined by the average kinetic energy of the individual particles.

The fundamentals of cold, low-pressure plasmas has been discussed in an extensive detail in the book *"Cold Plasma in materials fabrication" by Alfred Grill, IBM Research Division, T.J Watson Research Center, New York* [10]. The behavior of plasma closely resembles the behavior of neutral gas, and it is possible to somewhat explain its property as described by *Kinetic Theory of Gases.* In contrast to neutral gas molecules, the movement of particles composing plasma results in localized concentrations of positive and negative charges. These concentrations of charge give rise to long-range coulombic fields that influence the trajectories of charged particles. Consequently, a charged particle in plasma tends to follow a path that, on average, aligns with the electric field present.

Plasma is broadly characterized by the following basic parameters,

- The densities of electrons & ions: $n_e = n_i =$ which represents the plasma density.
- The density of neutral particles (n_n)
- The energy distributions of neutrals, f_n (w), ions, f_i (w) and electrons f_e (w)
- Electron Energy (T_e)
- Debye length λ_D which serves as a characteristic dimension indicating regions where the breakdown of neutrality can occur.

$$T_e = 1\,\text{eV},\, n_e = 1010\,\text{cm}^{-3},\, \lambda_D = \sqrt{\frac{\varepsilon_o k T_e}{n_e e^2}} = 74\,\mu\text{m} \quad (1)$$

- The presence of a plasma sheath near a surface is a result of electrons moving at significantly greater speeds than ions, causing them to depart from the plasma and create a positive charge in the surrounding area of the surface.

Near the surface, an electric field emerges that hinders the motion of electrons and accelerates ions. As a result, the surface acquires a negative potential, referred to as a negative self-bias, relative to plasma. In contrast, the plasma itself maintains a positive potential. The localized concentration of charged particles is confined within a volume equivalent to the Debye length (λ_D). It is important to highlight that beyond this volume, the charge density of ions matches the charge density of electrons, leading to the plasma's overall electrical neutrality. This property is the key reason why plasma is commonly regarded as a quasi-neutral gas.

1.2 Cold-plasma generation

To create and maintain cold plasma, gaseous molecules are typically exposed to direct current (DC), radio frequency (RF), or microwave (MW) power. In 1923, Irving Langmuir discovered the distinctive characteristics of DC glow discharge in a discharge tube and coined the term "plasma." It was observed that when voltage is applied to the gas in the discharge tube, the number of free electrons gradually increases and accelerates in response to the applied electric field, leading to a steady rise in the current (I). Once the applied voltage reaches a certain threshold value, known as the breakdown potential (V_b), an avalanche process occurs due to the interplay of three simultaneous processes:

- Accelerated ions collide with the cathode, resulting in the emission of secondary electrons.

- The newly generated ions, in turn, produce additional electrons through collisions as they accelerate toward the cathode, leading to the creation of more ions.

- The electrons produced in this manner are either removed from the plasma through recombination with oppositely charged ions or diffuse/drift toward the walls of the tube.

When equilibrium is reached between rate of formation and recombination of electrons, the discharge becomes self-sustaining. This leads to a significant breakdown in the gas, causing it to emit a glowing effect. Consequently, there is an abrupt rise in current accompanied by a voltage drop. The following schematics (**Figure 1**) summarize what prevails in plasma reactors commonly used in most laboratories.

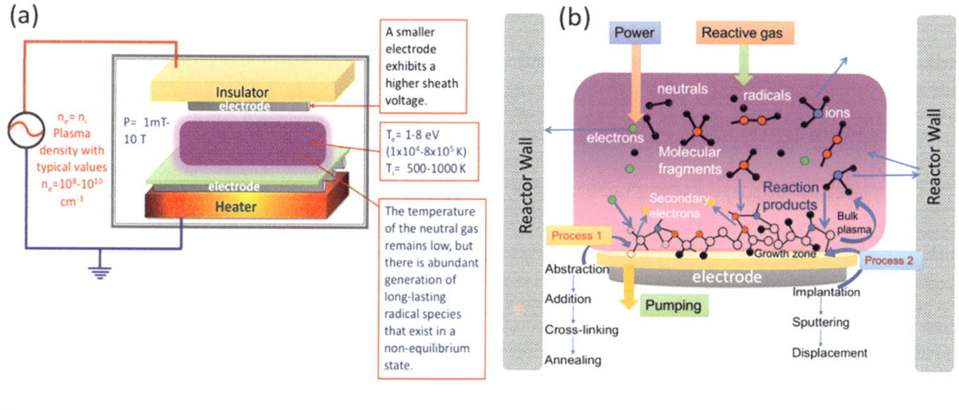

Figure 1.
Schematic depicting typical processes inside the plasma reactor.

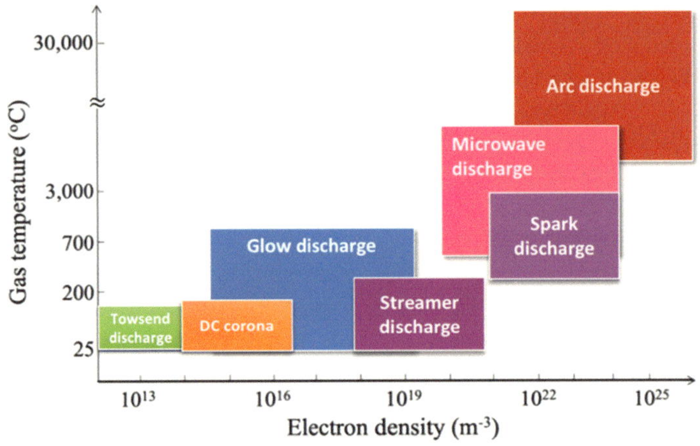

Figure 2.
Variation of electron density with gas temperature for gas discharges of the typical atmospheric cold plasma. (Adapted from ref. [5].)

In recent studies, cold plasmas have gained significant attention and are generated using methods such as corona discharges, dielectric barrier discharges, and plasma jets. These plasmas are created at atmospheric pressure by employing pulses with durations ranging from 10^{-6} to 10^{-9} seconds. In this type of plasma, known as 'Cold Atmospheric Plasma' (CAP), electrons with high energy levels are generated. CAP has shown great potential for a wide range of surface engineering application [12] (**Figure 2**).

2. Plasma assisted surface modification

The plasma-based surface modification majorly involves: *Plasma treatment*, *plasma etching*, and *plasma polymerization*. In case of plasma treatment of materials as well as plasma polymerization on a substrate, the plasma can be majorly divided into following three categories [13]:

- Chemically non-reactive plasma i.e. plasma generated by ionizing inert gas such as Ar, He, etc.
- Chemically reactive plasma i.e. plasma generated through inorganic and organic molecular gases such as O_2, N_2, CF_4, etc.
- Polymer forming plasma i.e. plasma resulting from organic or inorganic vapor.

Every year the substantial increase in the number of research papers and patents on the use of plasma technology for producing a variety of new materials and functional coatings is indicating the myriad of opportunities created by plasma technology [14]. Plasma-based methods are highly prevalent in surface treatment technologies and find extensive application in areas such as cleaning, surface activation, enhancing adhesion, corrosion resistance, biomedical applications, and the development of self-cleaning coatings [14–20].

Plasma techniques are employed in various processes related to substrate treatment, including cleaning the substrate using gaseous plasma prior to coating, as well as modifying the surface after applying the desired coating material [21, 22]. Plasma etching also known as dry etching, has emerged as a crucial technique in microfabrication processes for microelectronic devices, aligning with advancements in the field of microelectronics [10]. However, so far plasma-based approaches were not extensively used to develop self-cleaning surfaces in commercial scale. Vacuum based plasma techniques impose major challenges with respect to cost-effective large area deposition in ambient conditions. Though, one can find in several published literatures the use of plasma deposition techniques for developing functional self-cleaning coatings.

2.1 Plasma cleaning

Plasma treatment is a highly effective method for preparing materials for subsequent processing, as it enhances their surface energy and removes contaminants introduced during the production process. This treatment significantly improves the quality, durability, and adhesive properties of coatings or printing applied to the surface. By subjecting materials to plasma treatment, their surface is modified in a way that optimizes their compatibility with secondary manufacturing applications [23]. The process involves a reactive treatment wherein positive and negative ions, electrons, and radicals interact and collide in the presence of an electric potential difference. Through this energetic interaction, plasma treatment primes the surface, creating an environment that enhances the acceptance and bonding of subsequent manufacturing processes. By increasing the surface energy and eliminating contaminants, plasma treatment ensures that adhere more effectively, resulting in superior quality and longer lifespan of the applied materials. *Roger L. Shannon* and *Roger B. Gillette* together filed a patent that claims a plasma cleaning device for cleaning contaminated surfaces present in a high vacuum [24]. Basic steps involved in plasma cleaning of a substrate have been presented in **Figure 3**.

2.2 Plasma etching

Extensive research has been conducted on plasma-assisted surface modification techniques, aimed at customizing various surface properties through the treatment of target substrates with gaseous plasma. The etching process facilitated by chemically reactive plasma involves a combination of physical and chemical mechanisms, which depend on the stability and volatility of the products formed through the interaction between the etching gas and the material [13]. Etching by chemically non-reactive plasma occurs by a physical deposition process such as magnetron sputtering [25, 26].

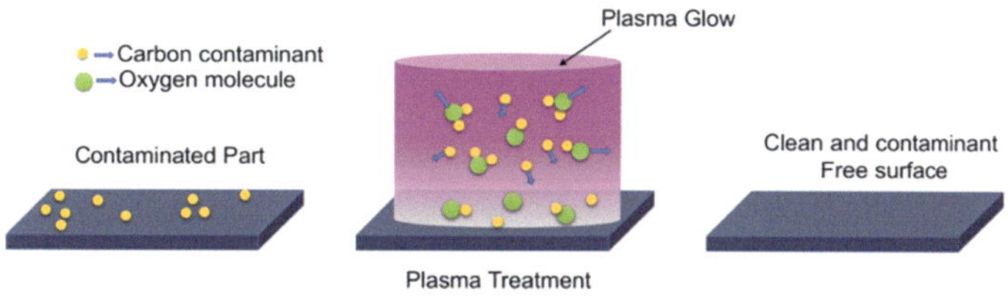

Figure 3.
Steps involved in typical plasma cleaning process.

The plasma etching is carried using a gaseous discharge containing reactive species such as atoms and radicals that react with the surface to form volatile compounds that evaporate from the surface under proper reaction condition leaving an etched target substrate. It is also possible to create desired nanostructure on the substrate by varying the process parameters and underlying plasma chemistry of the reactive etchant gas [27, 28].

The reactive ion etching (RIE) is one of the most widely used dry etching techniques. The etching in this system is achieved by suitable combination of chemical reactivity of plasma species and the physical effects caused by ion bombardment [29]. A schematic depiction of a typical RIE system is given **Figure 4**.

It should be noted that in RIE systems both the chemical reaction on the surface of target substrate and the removal of the volatile products so formed on the etched surface are enhanced by ion bombardment of gaseous plasma. Basic steps involved in plasma cleaning of a substrate have been presented in **Figure 5**.

2.3 Plasma polymerization

Plasma polymerization is a process that involves the deposition of an organic polymer through plasma-enhanced chemical vapor deposition (PECVD). This process entails the decomposition of an organic monomer gas, followed by the deposition and polymerization of the excited species onto the substrate's surface. Films deposited through plasma polymerization can have thicknesses ranging from several

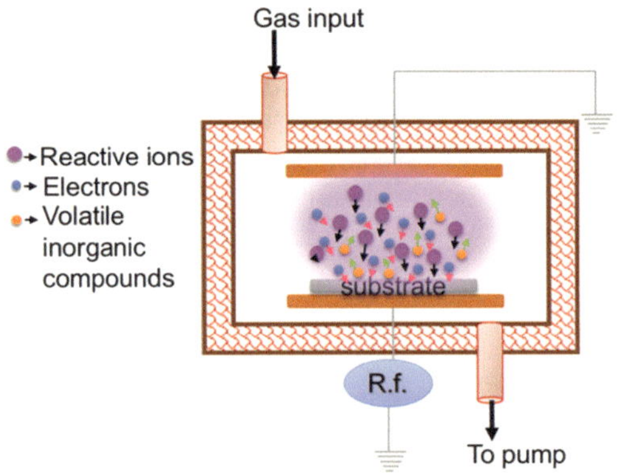

Figure 4.
Schematic depicting dry etching in a reactive ion etching system.

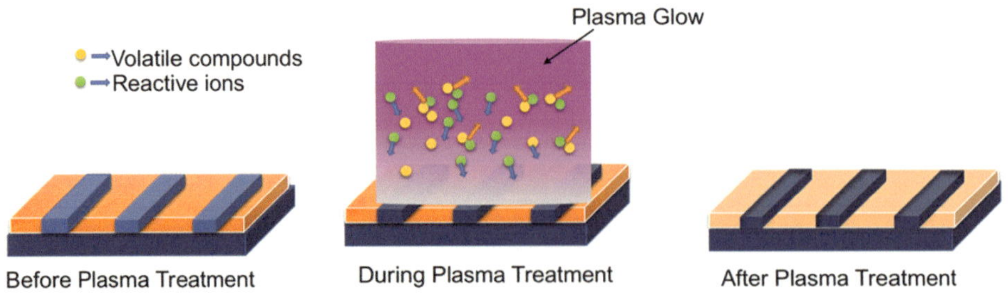

Figure 5.
Steps involved in a typical plasma assisted etching.

tens to several thousands of angstroms. These deposited films, known as plasma polymers, exhibit chemical and physical properties that differ from conventional polymers. Starting materials for plasma polymerization can include hydrocarbons, fluorocarbons, and organic compounds containing nitrogen, oxygen, or silicon. Plasma polymers do not possess distinct repeating units like conventional polymers. Instead, their properties are primarily determined by the plasma parameters used during the deposition process. For example, it is not possible to identify a specific product as a plasma polymer of ethylene, as various products can be obtained from an ethylene plasma. Therefore, plasma polymers derived from ethylene are not equivalent to polyethylene. It's worth noting that the presence of functional groups, such as double bonds, is not a prerequisite for the monomer used in plasma polymerization. The process can occur even without such functional groups. Furthermore, the process of plasma polymerization yields a polymer coating through a rapid step growth reaction mechanism [13]. Plasma polymerization does not rely on a single species for its occurrence; instead, a diverse range of active species generated during the process actively participate in the polymerization reaction. The specific type and quantity of these activated species are contingent upon the reaction conditions employed during plasma polymerization [30]. Since both the kind of reactive species (monoradical and diradicals) are formed simultaneously, reaction between them is also very much likely to occur by cross cycle reactions, depending on the concentration of both the species.

Important factors that determines the deposition of films by plasma polymerization processes are type of monomer, carrier/reactive gas and it mass flow, pressure, temperature, input power, and reactor type [31]. Another important factor that ascertains the several important surface properties of plasma assisted modification is the frequency source used to generate plasma.

Srivatsa et al. developed a custom plasma polymerization deposition system (**Figure 6**) to enhance both surface wettability and hardness [24]. Subsequently, a four-layer antireflective coating (ARC) is applied in the same chamber on a polycarbonate (PC) substrate, consisting of alternating thin films of titanium dioxide and

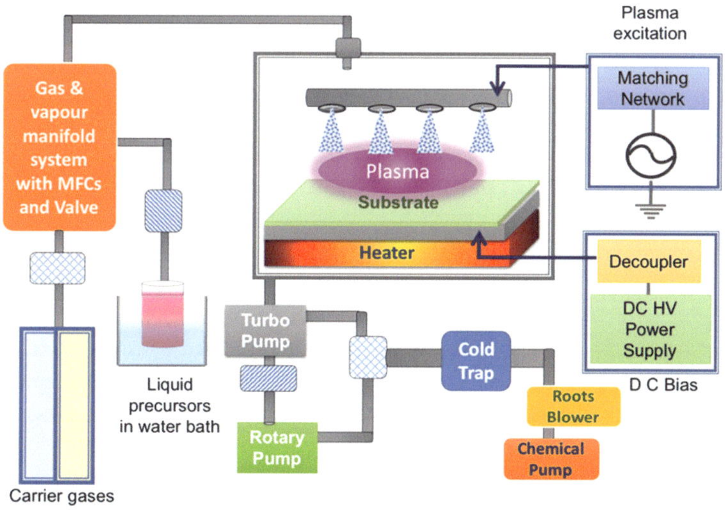

Figure 6.
Typical plasma polymerization deposition system set up. (Reproduced from ref. [24].)

silicon dioxide with varying thicknesses. This results in the creation of a robust and durable ARC on the PC substrate.

Jung *et al.* demonstrated effect of microwave and rf plasma on the photocatalytic properties and wetting behavior of TiO_2 thin films deposited on glass substrate using dip-coating techniques [32]. Han *et al.* adopted the plasma polymerization processing method for obtaining a broadband transparent, superhydrophobic surfaces usingCH_4, C_4F_8, and He gas mixture in an Rf plasma reactor at atmospheric pressure [33]. Ji et al. conducted a study where they presented the production of a superhydrophobic coating using a combination of atmospheric pressure middle frequency (Mf) and radio frequency (Rf) plasmas in an in-line process [34]. The coating was created by utilizing a mixture of Ar gas and HMDSO solution.

Barankin et al. developed a novel low-temperature atmospheric pressure plasma technique for polymerizing liquid fluoroalkylsilane (FAS) precursors, providing an effective method for producing hydrophobic coatings on glass and acrylic surfaces [35]. This innovative approach involves the atmospheric plasma curing of liquid FAS precursors, resulting in films with exceptional properties. The coatings exhibited a total surface energy of 11 dynes/cm, characterized by 10% polarity and average water contact angles ranging from 106° to 110°. To enhance adhesion between the FAS molecules and the surfaces, oxygen plasma activation was utilized, proving crucial in achieving strong bonding. The hydrophobic coatings generated through this method exhibit excellent transparency, stability, and long-lasting performance for over a year. This durability is attributed to the robust bonding between the fluoroalkyl ligands and the surfaces, preventing degradation and ensuring sustained hydrophobic characteristics.

2.4 Atmospheric pressure plasma

Various plasma sources (ion sources) which worked under vacuum conditions have long been used to tailor material properties [36]. In a major development in recent years their counter parts for processing under atmospheric conditions have started becoming available.

To enable affordable access to plasma techniques, researchers are now extensively exploring atmospheric pressure-based methods. There is a wide variety of atmospheric-pressure plasmas used in the processing of materials. Traditional sources of plasma, such as transferred arcs, plasma torches, corona discharges, and dielectric barrier discharges, are being employed. These sources generate plasmas with electron and neutral temperatures exceeding 5000°C, and the density of charged particles generally ranges from 1016 to 1019 cm^{-3} [37, 38]. They are primarily used in metallurgy applications due to their high-temperature nature. On the other hand, non-equilibrium plasmas created by corona discharges and dielectric barrier discharges have gas temperatures ranging from 50 to 400°C, and the density of charged particles falls within the range of 1010 to 1015 cm^{-3} [39, 40]. In order to create a dielectric barrier discharge (DBD) plasma, a high-frequency voltage is applied to a ceramic material, which leads to the generation of plasma within the ceramic's holes. Consequently, objects like an endless fabric tape can be passed over the ceramic, allowing for a significant surface area to be activated in a short amount of time. It is important to note that DBD plasma production results in the generation of ozone (O_3) [41]. Therefore, it is crucial to utilize an extraction system when operating DBD for extended periods in enclosed spaces. Popular atmospheric plasma deposition has been presented in **Figure** 7.

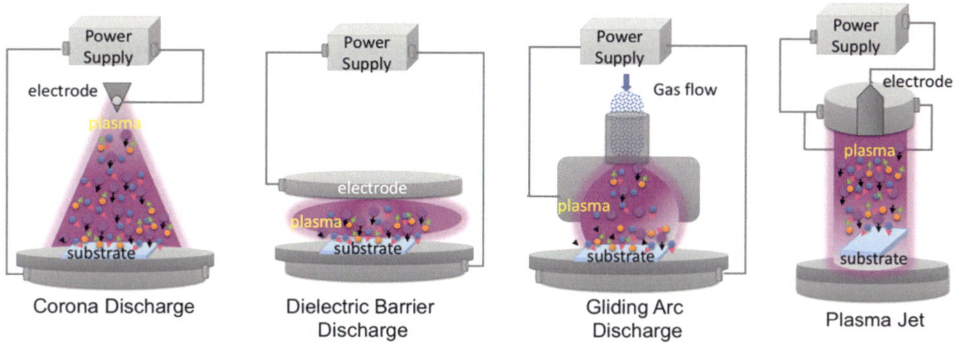

Figure 7.
Various atmospheric plasma deposition processes.

Among various atmospheric plasma processes, DBD plasma technique is highly advantageous due to its ability to generate high-energy electrons and facilitate rapid reactions. Moreover, it can be employed with various gases without causing electrode corrosion. As a result, this technique has found wide applications in diverse fields, including sterilization, material surface cleaning, semiconductor manufacturing, and surface treatment. Owing to its versatility and effectiveness, further research can explore its potential for optimizing deposition parameters and investigating the long-term stability and durability of plasma-treated surfaces in different environmental conditions. However, their non-uniform nature limits their use in materials processing.

In contrast, the atmospheric-pressure plasma jet shares similarities with a conventional low-pressure glow discharge [40, 42–44]. The plasma jet typically maintains a gas temperature between 25 and 200°C, with charged-particle densities at approximately 1011 cm^{-3}. Additionally, there are high concentrations of reactive species, ranging from 10 to 100 ppm. An important advantage of this plasma source is that it can operate without requiring a vacuum chamber, which significantly broadens its range of potential applications. In summary, atmospheric-pressure plasma jets present a promising solution for a wide range of materials processing applications. They offer moderate gas temperatures, relatively high particle densities, and a rich presence of reactive species. The absence of a vacuum requirement enhances their versatility, allowing them to be effectively utilized in various material processing scenarios. Moreover, to address the constraint imposed by the reactor size, which limits the use of moderate-sized substrates for functionalization or modification, a recent technology called air plasma spray has emerged. This technology aims to enhance the adhesion, uniformity, mechanical hardness, and other surface properties of the target substrate.

In a recent study by Marchand et al., they showcased the preparation of long-lasting superhydrophobic films on various substrates under ambient conditions, eliminating the need for a vacuum chamber. They achieved this by utilizing a tetramethylsilane (TMS) precursor [45]. A superhydrophobic coating was successfully demonstrated by Jin et al. using an in-line process that incorporated atmospheric pressure middle frequency (Mf) and radio frequency (Rf) plasmas [34]. The process involved the utilization of a mixture of Ar gas and HMDSO solution. Han et al. introduced a single-step processing technique for achieving broadband transparency (covering wavelengths from visible to near-infrared range) and super-hydrophobicity on surfaces. This was accomplished using a mixture of helium (He), methane (CH_4), and octafluorocyclobutane (C_4F_8), gases in a radio frequency (RF) plasma reactor operating at atmospheric pressure [33]. By employing atmospheric pressure, He/CH_4 plasma, it is possible to achieve a chemically dominant hydrophobicity without the need for etching agents. The effects of plasma treatment on hydrophobicity and transparency were investigated and a large contact angle ~165° and low hysteresis of 4.9° was obtained.

Hossain et al. investigated tetramethylsilane (TMS) and 3-aminopropyl methyl silane (APDMES) as precursors to apply superhydrophobic coatings onto glass substrates. The deposition process involved a plasma jet that relied on dielectric barrier discharge [46]. The resulting films demonstrated exceptional water repellency properties, with a water contact angle of 163° and a sliding angle of only 5°. Additionally, the films exhibited a surface roughness of 50 nm, which further enhanced their superhydrophobic characteristics and self-cleaning abilities. The deposition process utilizing the dielectric barrier discharge plasma jet proved to be a straightforward and efficient technique for creating superhydrophobic coatings on various substrates.

In a separate study, Lin et al. utilized a CO/N_2 DBD plasma along with acrylic acid as a precursor to alter the surface of a PTFE film [47]. This process led to creation of hydrophilic functional groups such as –COOH, C–O, C–C, and others on the PTFE surface. Through two reaction treatments, the water contact angle decreased to 140°, and after twenty reactions, it reached less than 5°, indicating a significant improvement in surface wettability.

Fakhouri et al. introduces a novel approach for depositing TiO_2 coatings using atmospheric pressure plasma jet (APPJ) coupled with a liquid precursor spray (droplet size~100 μm) [48]. The developed technique enables high deposition rates, and the resulting coatings exhibit a highly porous structure with pore sizes ranging from 100 to 1000 nm, leading to a large specific surface area. The coatings deposited by APPJ demonstrate several notable advantages over the RF-sputtered coatings. Firstly, they exhibit a significantly higher specific surface area, which is beneficial for enhanced photocatalytic activity. The increased surface area provides more active sites for reactions, leading to improved catalytic performance. Moreover, the porous nature of the APPJ coatings further contributes to their enhanced specific surface area and catalytic activity. In comparison to RF sputtering, the APPJ technique offers superior deposition properties, including higher deposition rates and improved coating characteristics. These findings highlight the effectiveness of the APPJ technique in depositing TiO_2 coatings with high specific surface areas and improved photocatalytic activity compared to RF-sputtered coatings.

3. Plasma assisted self-cleaning surface

Plasma coatings represent an exciting field within plasma technology, with significant potential for enhancing the functionality and value of materials across a wide range of applications. These coatings provide two primary surface properties: complete repellency to liquids (such as water and oil) or complete wetting ability. The process involves the formation of a nanoscale polymer layer across the entire surface of an object when it is exposed to plasma. This coating process is rapid, typically taking only a few minutes to complete. It is a very thin (few nanometers to microns) permanent coating, firmly bound to the material surface on an atomic scale.

Plasma coating technology generates surfaces with exceptional hydrophobic properties, exhibiting a "lotus effect" where water is repelled. These surfaces also possess dirt-repellent characteristics, enabling self-cleaning without the need for manual cleaning methods. Promising applications for plasma coatings include the coating of automotive components like aluminum wheel rims, as well as window frontages and glass panes [49].

Song *et al.* reported a one-step nanostructure fabrication process using plasma-assisted reactive ion etching (RIE) processing [50]. Mazumder *et al.* reported a similar hierarchical nanostructured transparent antireflection glass material with superomniphobic properties. The two-tier roughness was found to consist of primary nanopillars (~100–200 nm) superimposed with nanoparticles (~10–30 nm). The nanopillars were fabricated by metal dewetting followed by RIE and the nanoparticles are deposited using combustion chemical vapor deposition (CCVD). They have been able to achieve contact angles ~170° and 160° for water and oil, respectively and an average transmission of 93.8% in the wavelength range 400 to 700 nm using CF_4 and O_2 gas mixture, exhibits very high optical transmission and antifogging effects [51].

3.1 Versatility of low temperature plasma

In recent times, the advancement of low temperature plasma technology has emerged as a flexible and efficient approach for manipulating the surface structure of materials [52]. One specific technique, known as low temperature plasma-enhanced chemical vapor deposition (PECVD), offers several notable advantages. These include direct surface pretreatment, modification at low temperatures, low organic content, and the absence of post-processing requirements [53, 54]. PECVD utilizes the energy and reactivity of electrons, ions, free atoms, and free radicals present in the plasma to induce physical and chemical transformations on the material surface. Consequently, this process enables the formation of new functional groups with desirable properties or the creation of film layers with unique structures, making it a crucial method in the treatment of superwetting surfaces.

The significance of plasma-based process to achieve surface superwettability is evident from the surge in research publications. In the year 2000, the number of papers published on this topic was less than 3500. However, by 2010, this number had increased to nearly 9000, highlighting the expanding interest in the field. As of the current year, approximately 20,000 papers have been published in the area of plasma control of surface superwettability, demonstrating its continued relevance and widespread research focus [55].

Two widely used approaches for surface modification are atmospheric pressure plasma treatment and low-pressure plasma treatment [56, 57]. Atmospheric pressure plasma treatment includes techniques like dielectric barrier discharge plasma and atmospheric pressure plasma jet treatment. On the other hand, low-pressure plasma treatment involves methods such as radiofrequency (Rf) discharge plasma treatment and glow discharge plasma treatment. These methods enable the application of plasma to the surface, leading to the creation of nano-scale roughness [58]. Alternatively, by introducing precursors that are chemically linked to diverse functional groups, the desired surface wettability can be achieved [59].

Zhang et al. explored the potential of an atmospheric pressure air plasma jet, generated using a dielectric barrier structure with hollow electrodes (HEDBS), as a means to spray TiO_2 films inside tubular substrates [60]. The results obtained from the self-cleaning test indicate that the proposed HEDBS approach, utilizing air as the working gas, offers a practical and efficient method for synthesizing thin TiO_2 nanofilms, while simultaneously reducing costs and saving time. This approach holds significant promise for various applications, particularly for coating tubular substrates with TiO_2 films.

Chemin et al. presented innovative method for synthesizing anatase TiO_2/SiO_2 nanocomposite coatings at atmospheric pressure, offering transparent anti-fogging and self-cleaning properties on polymer substrates [61]. The concomitant injection of titania and silica precursors in a blown-arc discharge allows for the efficient production of durable coatings with excellent performance characteristics. The findings of this study pave the way for the development of advanced coatings applicable in various industries where transparency, anti-fogging, and self-cleaning properties are highly desirable. The use of anatase TiO_2 nanoparticles in conjunction with SiO_2 in the nanocomposite coatings contributes to their enhanced durability and desirable properties. The combination of these materials not only imparts super-hydrophilicity, which facilitates water spreading and evaporation, but also harnesses the photocatalytic nature of TiO_2, enabling the degradation of organic pollutants and further promoting self-cleaning functionality.

This plasma-based surface pre-treatment using air plasma gun results in clean and highly activated surfaces that can be easily wetted by water, eliminating the need for volatile organic compounds (VOCs) and the associated processes of drying and wastewater disposal. This technology finds application in various areas such as enhancing structural adhesive bonds in the automotive industry, creating seals for electronics applications, or achieving quick and bubble-free wet labelling with strong initial adhesion in the packaging industry. Moreover, the use of plasma gun has been reported for depositing thin film coatings like organosilicate and ceria nanocomposites, which provide improved hardness, adhesion, and UV-absorbing properties [62, 63].

Deicing and self-cleaning processes have gained significant global attention due to their crucial importance in various industries. One major challenge faced by these industries is the formation of atmospheric ice caused by super-cooled dewdrops in icy rain. This phenomenon poses a significant threat to aircraft, wind turbines, electric and telecommunication wires, highways, bridges, and other outdoor equipment and systems. Plasma-treated superhydrophobic (SH) coatings have emerged as a highly relevant solution for both deicing and self-cleaning applications in aircraft and other systems. Subeshan et al. the use of plasma-treated superhydrophobic coatings provides an effective solution for both deicing and self-cleaning requirements in aviation industries [64]. This study contributes valuable insights into the characteristics and performance of plasma-treated SH coatings on the AA 2024-T3 alloy, showcasing their ability to repel water, prevent ice formation, and maintain a clean surface under diverse atmospheric conditions.

The utilization of an air plasma gun for plasma-based surface pre-treatment leads to the formation of clean and highly activated surfaces that readily wet with water. This eliminates the need for volatile organic compounds (VOCs), as well as the associated processes of drying and wastewater disposal. The technology has diverse applications, including enhancing adhesive bonds in the automotive industry, creating seals for electronics, and achieving efficient and bubble-free wet labelling with strong initial adhesion in the packaging industry. Additionally, the plasma gun has been proven effective for depositing thin film coatings like organosilicate and ceria nanocomposites, which exhibit improved hardness, adhesion, and UV-absorbing properties [54, 55]. Hovish et al. presented a technique for growing a superhydrophilic organotitanate thin film with excellent thermomechanical properties using atmospheric plasma deposition. This thin film demonstrated an effective antifogging property, with a contact angle (θ) less than 5° [65]. Asadollahi et al. described the development of an anti-icing coating on Al-6061 sheets. This involved pre-treating the surface and depositing organo-silane (HMDSO) using an atmospheric pressure plasma jet (APPJ) [66].

Additionally, Plasma jet printing (PJP) is a highly versatile and scalable printing method that utilizes plasma to interact with aerosolized inks, offering precise control over material characteristics. PJP represents an innovative approach for high-quality material printing, as it is environmentally friendly and can utilize lower quality inks. The process involves transforming the material into aerosol form, introducing it into the print head, and generating plasma through an electromagnetic field. The reactive species within the plasma interact with the ink, allowing for dynamic in situ tailoring of electronic properties. PJP's gas-phase operation eliminates liquid waste and toxic by-products, while enabling the use of unprocessed, low-grade inks to produce high-quality prints [51].

3.2 Integration of sol–gel and gaseous plasma-based approaches

The combination of sol–gel and plasma-based techniques has led to the development of coatings with unique optical and surface properties. Jung et al. conducted a study that specifically explored surface modification using plasma treatment on a sol–gel derived TiO2 film, which served as a photocatalyst [32]. The study investigated the effect of gaseous plasma treatment (Ar/O_2) on the enhancement of wetting behavior, specifically the attainment of a superhydrophilic property. In one of our own research work, we investigated sol–gel derived V-TiO_2:SiO_2 coatings and applied subsequent O_2 plasma treatment [59]. Different weight percentages of V-doped TiO_2 in the SiO_2 sol were tested to achieve the best combination of optical transparency and self-cleaning properties. The resulting coatings demonstrated antireflective properties, superhydrophilicity, and visible photo-catalytic activity. O_2 plasma treatment significantly reduced the contact angle of all V-TiO_2:SiO_2 formulations, reaching 0°. In another study, we combined sol–gel self-assembly with a plasma-based approach to create highly transparent, self-ordered, superhydrophilic, and photoactive TiO_2 thin film coatings [60]. The TiO_2 sol contained a block copolymer that facilitated the formation of regular pores in the film, resulting in reduced refractive index values (~1.31) and improved transparency (4% antireflection gain). The mesoporous TiO_2 coatings exhibited excellent photocatalytic activity for VOCs. Nitrogen plasma treatment enhanced mechanical stability and hydrophilicity without affecting optical transmission. Under optimized conditions, the coatings achieved superhydrophilicity with a water contact angle of less than 5°. Optimal conditions were determined by varying the Rf self-bias potential and duration of plasma treatment.

Noppakun Sampo conducted research on the utilization of sol–gel methods in the development of a multifunctional magnetic iron-based solution for thermal spray techniques [61]. The study focuses on preparing liquid feedstocks for the solution precursor plasma spray (SPPS) process. By exploring sol–gel methods in thermal spray applications, valuable insights were gained into the development and characterization of the magnetic iron-based solution. The research specifically investigates the effects of organic chelating agents on the topography, physical properties, and phase composition of cobalt ferrite splats. These findings contribute to a deeper understanding of the mechanisms involved and have significant potential for practical applications in various fields. Similarly by employing the SPPS technique, Cai et al. successfully developed ceramic superhydrophobic coatings with desirable properties [67]. The deposition process has been shown in **Figure 8**. The chosen rare earth oxide as the

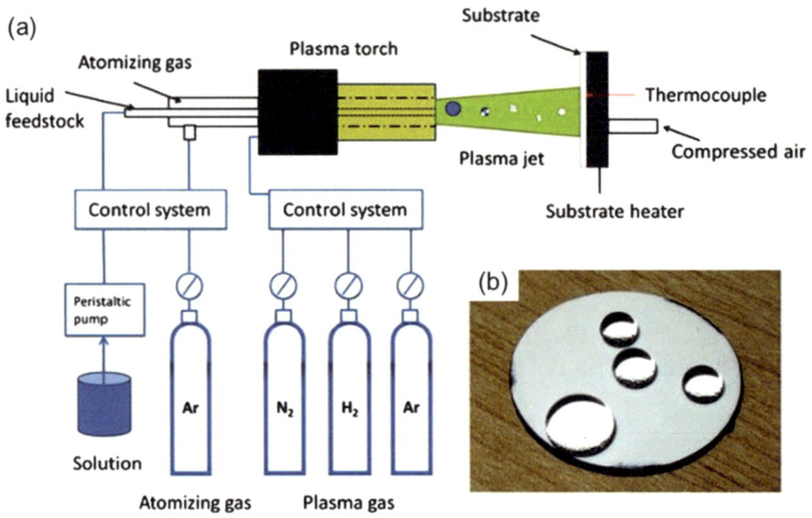

Figure 8.
(a) Schematic of the SPPS deposition system. (b) Water droplets of varying sizes on the coated surface (with permission, Copyright © 2016 Yuxuan Cai et al.).

coating material, along with the optimized spraying conditions, resulted in coatings with excellent hydrophobicity, mimicking the natural superhydrophobic surfaces. The outcomes of this study have significant implications for various applications where enhanced water repellency and self-cleaning properties are desirable, such as anti-icing coatings, corrosion protection, and efficient fluid flow management.

The sol–gel process is a commonly used method for synthesizing powders, coatings, and bulk materials. However, as a wet chemical technique, it has certain limitations, especially when applied as a coating method using aqueous colloidal solutions. The commonly utilized techniques such as dip and spin coating face challenges when applied to complex substrates. To overcome these limitations, the aerosol gel deposition method provides a promising solution. Pietrzyk et al. reported a novel technique to deposit Al_2O_3 coating by plasma enhanced aerosol-gel method [68]. By combining the aerosol-gel deposition of thin films with low-temperature plasma treatment, an innovative approach is established, introducing a novel method known as plasma-enhanced aerosol-gel for coatings production. This combined approach brings forth numerous advantages. Firstly, it enables the application of coatings on more complex substrates, which are difficult to coat using traditional techniques. Secondly, the aerosol-gel deposition method provides a homogeneous and controlled deposition of thin films. Lastly, the subsequent low-temperature plasma treatment enhances the properties of the deposited coatings, such as adhesion, durability, and functionality. This integration of aerosol-gel deposition and low-temperature plasma treatment represents a significant advancement in the field of coatings production, providing a versatile and efficient approach for the synthesis of high-quality coatings with improved properties.

4. Conclusion

Gaseous plasma, also known as cold plasma, exhibits an ion/electron density of around 1000 cm^{-3} and an electron temperature of 1 eV. Different types of plasma reactors, such as DC, Rf, VHf, and MW, with varying excitation frequencies, play a crucial role in film processing. Pulsing the discharge subtly affects the film's characteristics.

Cold plasma finds applications in surface cleaning, etching, deposition, and grafting during film growth. Plasma polymerization overcomes limitations imposed by plasma chemistry, as plasma polymers' properties depend on plasma parameters rather than the monomer used. Surface superwettability control using gaseous plasma treatment (e.g., Ar/O_2) has been extensively explored, resulting in improved superhydrophilic properties.

Recent advancements allow the generation of cold plasma at atmospheric pressure using techniques like Corona, DBD, Gliding Arc discharge, and Plasma jet, overcoming the need for vacuum-based reactors. The atmospheric pressure plasma jet (APPJ) technique proves effective in depositing TiO_2 coatings with high specific surface areas and improved photocatalytic activity compared to vacuum-based plasma reactors. These approaches enable surface bombardment with plasma, creating nano-scale roughness. By introducing functionalized precursors, desired surface wettability can be achieved. Open-air plasma deposition is used for various thin film coatings, offering enhanced hardness, adhesion, and UV-absorbing properties. Plasma-treated superhydrophobic (SH) coatings are highly relevant for de-icing and self-cleaning applications in aircraft and other systems. An innovative method demonstrates the synthesis of anatase TiO_2/SiO_2 nanocomposite coatings at atmospheric pressure, providing transparent anti-fogging and self-cleaning properties on polymer substrates. Additionally, an atmospheric pressure air plasma jet generated using a dielectric barrier structure with hollow electrodes (HEDBS) shows promise for spraying TiO_2 films inside tubular substrates.

To overcome challenges with complex substrates, traditional techniques like dip and spin coating are supplemented by the aerosol gel deposition method. This approach integrates aerosol-gel deposition with low-temperature plasma treatment, enabling the novel plasma-enhanced aerosol-gel method for coatings production. Moreover, plasma jet printing offers a versatile and scalable printing method that leverages plasma's interaction with aerosolized inks, allowing precise control and customization of material characteristics.

These advancements in plasma technology offer exciting possibilities for self-cleaning applications. Plasma treatments can modify surface properties to achieve superhydrophilic or superhydrophobic behavior. The integration of plasma techniques with innovative deposition methods paves the way for improved coating performance and functionality on various substrates.

Author details

Deepanjana Adak[1*] and Raghunath Bhattacharyya[2]

1 Surface Engineering Division, CSIR-National Aerospace Laboratories, Bangalore, India

2 School of Advanced Materials, Green Energy and Sensor Systems, Indian Institute of Engineering Science and Technology, Shibpur, India

*Address all correspondence to: deepanjana.adak@gmail.com

References

[1] John PI. Plasma Sciences and the Creation of Wealth. Tata McGraw-Hill Publishing Company; 2005. ISBN: 9780070585577

[2] Lee DH, Song Y-H, Kim K-T, Jo S, Kang H. Current state and perspectives of plasma applications for catalyst regeneration. Catalysis Today. 2019;**337**:15-27

[3] Pham VP, Jo YW, Oh JS, Kim SM, Park JW, Kim SH, et al. Effect of plasma–nitric acid treatment on the electrical conductivity of flexible transparent conductive films. Japanese Journal of Applied Physics. 2013;**52**(7R):075102

[4] Pham VP, Kim KH, Jeon MH, Lee SH, Kim KN, Yeom GY. Low damage pre-doping on CVD graphene/Cu using a chlorine inductively coupled plasma. Carbon. 2015;**95**:664-671

[5] Kim KN, Pham VP, Yeom GY. Chlorine radical doping of a few layer graphene with low damage. ECS Journal of Solid State Science and Technology. 2015;**4**(6):N5095

[6] Pham VP, Kim DS, Kim KS, Park JW, Yang KC, Lee SH, et al. Low energy BCl3 plasma doping of few-layer graphene. Science of Advanced Materials. 2016;**8**(4):884-890

[7] Pham VP, Kim KN, Jeon MH, Kim KS, Yeom GY. Cyclic chlorine trap-doping for transparent, conductive, thermally stable and damage-free graphene. Nanoscale. 2014;**6**(24):15301-15308

[8] Pham PV. A library of doped-graphene images via transmission electron microscopy. C. 2018;**4**(2):34

[9] Pham P et al., editors. 21st Century Surface Science: A Handbook. IntechOpen; 2020. DOI: 10.5772/intechopen.87891

[10] Grill A. Cold Plasma in Materials Fabrication. New York: IEEE Press; 1994. p. 272

[11] Tendero C, Tixier C, Tristant P, Desmaison J, Leprince P. Atmospheric pressure plasmas: A review. Spectrochimica Acta Part B: Atomic Spectroscopy. 2006;**61**(1):2-30

[12] Bruggeman PJ, Iza F, Brandenburg R. Foundations of atmospheric pressure non-equilibrium plasmas. Plasma Sources Science and Technology. 2017;**26**(12):123002

[13] Yasuda H. Plasma Polymerization. Elsevier Science; 1985. ISBN: 9780127687605

[14] Jafari R, Asadollahi S, Farzaneh M. Applications of plasma technology in development of superhydrophobic surfaces. Plasma Chemistry and Plasma Processing. 2013;**33**(1):177-200

[15] Isabell TC, Fischione PE, O'Keefe C, Guruz MU, Dravid VP. Plasma cleaning and its applications for electron microscopy. Microscopy and Microanalysis. 1999;**5**(2):126-135

[16] Carosio F, Alongi J, Frache A. Influence of surface activation by plasma and nanoparticle adsorption on the morphology, thermal stability and combustion behavior of PET fabrics. European Polymer Journal. 2011;**47**(5):893-902

[17] Liston EM. Plasma treatment for improved bonding: A review. The Journal of Adhesion. 1989;**30**(1-4):199-218

[18] Friedrich JF, Mix R, Kühn G. Adhesion of metals to plasma-induced functional groups at polymer surfaces. Surface and Coatings Technology. 2005;**200**(1):565-568

[19] Singh H, Sidhu BS, Puri D, Prakash S. Use of plasma spray technology for deposition of high temperature oxidation/corrosion resistant coatings: A review. Materials and Corrosion. 2007;**58**(2):92-102

[20] Chu PK, Chen JY, Wang LP, Huang N. Plasma-surface modification of biomaterials. Materials Science and Engineering: R: Reports. 2002;**36**(5):143-206

[21] d'Agostino R, Favia P, Kawai Y, Ikegami H, Sato N, Arefi-Khonsari F, editors. Advanced Plasma Technology. John Wiley & Sons; 2007. DOI: 10.1002/9783527622184

[22] Adak D, Bhattacharyya R, Barshilia HC. A state-of-the-art review on the multifunctional self-cleaning nanostructured coatings for PV panels, CSP mirrors and related solar devices. Renewable and Sustainable Energy Reviews. 2022;**159**:112145

[23] Pham PV. Cleaning of graphene surfaces by low-pressure air plasma. Royal Society Open Science. 2018;**5**(5):172395

[24] Shannon RL, Gillette RB, inventors; US Patent assignee. Plasma cleaning device. 1978

[25] Hossain MM. Plasma Technology for Deposition and Surface Modification. Logos Verlag Berlin GmbH; 2009. ISBN: 978-3-8325-2074-8

[26] Pham PV. The new etching technologies of graphene surfaces. In: 21st Century Surface Science: A Handbook. London, UK: IntechOpen; 2020. DOI: 10.5772/intechopen.92627

[27] Xu S, Levchenko I, Huang SY, Ostrikov K. Self-organized vertically aligned single-crystal silicon nanostructures with controlled shape and aspect ratio by reactive plasma etching. Applied Physics Letters. 2009;**95**(11):111505

[28] He B, Yang Y, Yuen MF, Chen XF, Lee CS, Zhang WJ. Vertical nanostructure arrays by plasma etching for applications in biology, energy, and electronics. Nano Today. 2013;**8**(3):265-289

[29] Bondur JA. Dry process technology (reactive ion etching). Journal of Vacuum Science and Technology. 1976;**13**(5):1023-1029

[30] Yasuda HK. Some important aspects of plasma polymerization. Plasma Processes and Polymers. 2005;**2**(4):293-304

[31] Hegemann D, Hossain MM, Körner E, Balazs DJ. Macroscopic description of plasma polymerization. Plasma Processes and Polymers. 2007;**4**(3):229-238

[32] Jung C-K, Bae IS, Song YH, Boo JH. Plasma surface modification of TiO_2 photocatalysts for improvement of catalytic efficiency. Surface and Coatings Technology. 2005;**200**(5):1320-1324

[33] Han D, Moon SY. Rapid formation of transparent Superhydrophobic film on glasses by $He/CH_4/C_4F_8$ plasma deposition at atmospheric pressure. Plasma Processes and Polymers. 2015;**12**(2):172-179

[34] Ji Y-Y, Hong Y-C, Lee S-H, Kim S-D, Kim S-S. Formation of super-hydrophobic and water-repellency surface with hexamethyldisiloxane (HMDSO) coating on polyethyleneteraphtalate

fiber by atmosperic pressure plasma polymerization. Surface and Coatings Technology. 2008;**202**(22):5663-5667

[35] Barankin MD, Gonzalez E, Habib SB, Gao L, Guschl PC, Hicks RF. Hydrophobic films by atmospheric plasma curing of spun-on liquid precursors. Langmuir. 2009;**25**(4):2495-2500

[36] Bhattacharyya R. Indigenous ion sources for material processing. Defence Science Journal. 2009;**59**(4)

[37] Kruger CH, Owano TG, Laux CO. Experimental investigation of atmospheric pressure nonequilibrium plasma chemistry. IEEE Transactions on Plasma Science. 1997;**25**(5):1042-1051

[38] Schutze A, Jeong JY, Babayan SE, Park J, Selwyn GS, Hicks RF. The atmospheric-pressure plasma jet: A review and comparison to other plasma sources. IEEE Transactions on Plasma Science. 1998;**26**(6):1685-1694

[39] Goldman M, Sigmond RS. Corona and insulation. IEEE Transactions on Electrical Insulation. 1982;**17**(2):90-105

[40] Eliasson B, Kogelschatz U. Nonequilibrium volume plasma chemical processing. IEEE Transactions on Plasma Science. 1991;**19**(6):1063-1077

[41] Dielectric Barrier Discharge. Available from: https://www.plasma.com/en/dielectric-barrier-discharge/diener

[42] Jeong JY, Babayan SE, Tu VJ, Park J, Henins I, Hicks RF, et al. Etching materials with an atmospheric-pressure plasma jet. Plasma Sources Science and Technology. 1998;**7**(3):282

[43] Herrmann HW, Henins I, Park J, Selwyn GS. Decontamination of chemical and biological warfare (CBW) agents using an atmospheric pressure plasma jet (APPJ). Physics of Plasmas. 1999;**6**(5):2284-2289

[44] Jeong JY, Park J, Henins I, Babayan SE, Tu VJ, Selwyn GS, et al. Reaction chemistry in the afterglow of an oxygen–helium, atmospheric-pressure plasma. The Journal of Physical Chemistry A. 2000;**104**(34):8027-8032

[45] Marchand DJ, Dilworth ZR, Stauffer RJ, Hsiao E, Kim J-H, Kang J-G, et al. Atmospheric rf plasma deposition of superhydrophobic coatings using tetramethylsilane precursor. Surface and Coatings Technology. 2013;**234**:14-20

[46] Hossain MM, Trinh QH, Nguyen DB, Sudhakaran MSP, Mok YS. Formation of plasma-polymerized superhydrophobic coating using an atmospheric-pressure plasma jet. Thin Solid Films. 2019;**675**:34-42

[47] Lin L, Rui L, Li C, Liu Q, Li S, Xia Y, et al. Study on CO_2-based plasmas for surface modification of polytetrafluoroethylene and the wettability effects. Journal of CO_2 Utilization. 2021;**53**:101752

[48] Fakhouri H, Salem DB, Carton O, Pulpytel J, Arefi-Khonsari F. Highly efficient photocatalytic TiO_2 coatings deposited by open air atmospheric pressure plasma jet with aerosolized TTIP precursor. Journal of Physics D: Applied Physics. 2014;**47**(26):265301

[49] Surface modification using PlasmaPlus® nanocoating. Available from: https://www.plasmatreat.com/en/what-is-plasma/plasma-surface-treatment/plasma-coating/functional-coating/plasmatreat

[50] Mazumder P, Jiang Y, Baker D, Carrilero A, Tulli D, Infante D, et al.

Superomniphobic, transparent, and antireflection surfaces based on hierarchical nanostructures. Nano Letters. 2014;**14**(8):4677-4681

[51] Song YM, Park GC, Kang EK, Yeo CI, Lee YT. Antireflective grassy surface on glass substrates with self-masked dry etching. Nanoscale Research Letters. 2013;**8**(1):505

[52] Lei Z, Xin C, Yang J, Zhang L, Hao S, Min L, et al. The denitration mechanism of fly ash catalysts prepared by low-temperature plasma technology. Vacuum. 2020;**181**:109695

[53] Kim J, Hwang JH, Kwon YW, Bae HW, An M, Lee W, et al. Hydrogen-assisted low-temperature plasma-enhanced chemical vapor deposition of thin film encapsulation layers for top-emission organic light-emitting diodes. Organic Electronics. 2021;**97**:106261

[54] Leszczynski S, Strobel C, Leszczynska B, Waurenschk S, Röhlecke S, Stahr F, et al. Dynamic deposition system for fabrication of amorphous/crystalline silicon heterojunction solar cells combining linear hot-wire and plasma enhanced chemical vapor deposition methods. Thin Solid Films. 2022;**754**:139296

[55] Li D, Xu K, Zhang Y. A review on research Progress in plasma-controlled superwetting surface structure and properties. Polymers. 2022;**14**(18):3759

[56] Haye E, Job N, Wang Y, Penninckx S, Stergiopoulos V, Tumanov N, et al. ZnO/carbon xerogel photocatalysts by low-pressure plasma treatment, the role of the carbon substrate and its plasma functionalization. Journal of Colloid and Interface Science. 2020;**570**:312-321

[57] Takenaka K, Machida R, Bono T, Jinda A, Toko S, Uchida G, et al. Development of a non-thermal atmospheric pressure plasma-assisted technology for the direct joining of metals with dissimilar materials. Journal of Manufacturing Processes. 2022;**75**:664-669

[58] Chang FJ, Nishijima D, Tynan GR. Rough-surface effect on sputtering of Cr bombarded by low-energy He plasma. Nuclear Materials and Energy. 2021;**29**:101077

[59] Ma X, Mao Z, Xu D, Ding Y, Xu C. High-rate synthesis of SiCN films using single-source silicon precursor with high-density helicon plasma. Vacuum. 2020;**177**:109397

[60] Zhang L, Yu S, Wang K, Zhang J, Fang J. A cost- and time-saving strategy of spraying TiO_2 self-cleaning coatings in tubular substrates by air cold plasma. Chemical Physics Letters. 2017;**687**:205-208

[61] Chemin J-B, Bulou S, Baba K, Fontaine C, Sindzingre T, Boscher ND, et al. Transparent anti-fogging and self-cleaning TiO_2/SiO_2 thin films on polymer substrates using atmospheric plasma. Scientific Reports. 2018;**8**(1):9603

[62] Cui L, Ranade AN, Matos MA, Pingree LS, Frot TJ, Dubois G, et al. Atmospheric plasma deposited dense silica coatings on plastics. ACS Applied Materials & Interfaces. 2012;**4**(12):6587-6598

[63] Ding Y, Dong S, Hilt F, Dauskardt RH. Open-air spray plasma deposited UV-absorbing nanocomposite coatings. Nanoscale. 2018;**10**(30):14525-14533

[64] Subeshan B, Usta A, Asmatulu R. Deicing and self-cleaning of plasma-treated superhydrophobic coatings on the surface of aluminum alloy sheets. Surfaces and Interfaces. 2020;**18**:100429

[65] Hovish MQ, Hilt F, Rolston N, Xiao Q, Dauskardt RH. Open air plasma deposition of Superhydrophilic Titania coatings. Advanced Functional Materials. 2019;**29**(19):1806421

[66] Asadollahi S, Farzaneh M, Stafford L. On the Icephobic behavior of organosilicon-based surface structures developed through atmospheric pressure plasma deposition in nitrogen plasma. Coatings. 2019;**2019**:9

[67] Cai Y, Coyle TW, Azimi G, Mostaghimi J. Superhydrophobic ceramic coatings by solution precursor plasma spray. Scientific Reports. 2016;**6**(1):24670

[68] Pietrzyk B, Miszczak S, Szymanowski H, Kucharski D. Plasma enhanced aerosol–gel deposition of Al_2O_3 coatings. Journal of the European Ceramic Society. 2013;**33**(12):2341-2346

Chapter 6

Nanomachining

K.S. Srin, J. Ramkumar and Ravi N. Bathe

Abstract

LIPSS (LASER Induced Periodic Surface Structures) is a term used to describe ordered or periodically structured nanostructures. Until newly, this term was not extensively researched or optimized for use in superhydrophobic self-cleaning applications. This machining method is one of the most sophisticated and cutting-edge ways to create nanofeatures like grooves and bumps without using any chemicals. Here we are trying to explore with polarization, machining speed, number of pulses, LASER fluence, shielding gas, and other parameters dependence on surface nanostructures and features that varied in size and orientation when milled by a LASER. Mainly, the aspect ratio of nano features are varied with axial spacing and along the horizontal diameter of the LASER beam. When polarization employed on nanomachining, the increased number of threshold LASER pulses also causes the structures to become LASER Induced Irregular Surface Structures, and the orientation and size of these features (between 200 and 400 nm) also affected by external interacts of material surface. The existing knowledge describes the nano feature generation is by the interaction of LASER beam and surface wavers. In light of this, one of the newest topics to emerge for the development of nanofeatures using femtosecond LASER.

Keywords: LIPSS, LASER materials processing, self-cleaning surfaces, femtosecond LASER, surface plasmon polaritons, polarization

1. Introduction

The concepts of LIPSS and laser-induced irregular surface structures need to be thoroughly investigated in order to justify their application at all levels of the pertinent femtosecond laser field. During the femtosecond laser machining, the surface molecules of the material are producing various sorts of self-assembled features ranging on few microns (200 nm - 400 nm) either in LIPSS or laser-induced irregular surface structures. The laser fluence's transition from near to far ablation threshold typically changes the morphology from a periodic configuration to a complicated configuration. The increase in laser fluence deepens the grooves and intensifies the roughness of its produced feature.

When a linearly polarized femtosecond laser beam strikes a solid under ambient conditions, parallel and perpendicular LIPSS with a spatial frequency close to the laser's wavelength are produced.

Early-career researchers have already discovered that the most significant influences on the development of LIPSS and laser-induced irregular surface structures are the processing environment, the laser beam's wavelength, angle of incidence, pulse frequency, and laser intensity. Another important consideration is the processing material's optical characteristics in the focused situation. On polished stainless steel, Stephan and Frank (2015) described the change of nano feature with linear polarization of a 300 fs laser at a wavelength of 1025 nm and the application is to optoelectronic and biological devices. He has done the experiment in the ambient condition and under controlled electric field vector and the outcome was LIPSS with a periodicity of 925 nm perpendicular to the electric field vector [1]. Picosecond laser with specific polarization on cold-worked tool steel for modified LIPSS manufacturing was also tested by Slovenian researchers Peter et al. in 2016. Additionally, the increment of the number of pulses deleting the previous direction of the ripples, as well as the generation of nano ripples perpendicular to the laser polarization, were observed [2]. Naoki et al. [3] provided evidence and proof of the material's invariability during the creation of LIPSS under femtosecond laser irradiation. The size and groove depth depend on processing material and its ability to harden, such as nitride steel or stainless steel, with any aspects that were constant. An ambient environment will offer spatially oriented surface structures in the laser wavelength range that are perpendicular to the polarization in 30 to 150 fs. When LIPSS was introduced, the optical properties of semiconductor and dielectric materials were changed, and it's possible that the pulse shaping will govern the LIPSS orientation [4]. By using femtosecond laser, it is possible to create grooves on a single crystal diamond that are even 40 nm in size. However, the aspect ratio features cannot be created until the laser energy is near to threshold. This approach can be suggested to develop the submicron feature in diamonds for various purposes [5].

In this study, we extrapolate the machining characteristics to standardize the fabrication parameters of LIPSS and laser-induced irregular surface structures through the laser parameter restrictions and working conditions functionalities. The complete assessment would be necessary to create accurate and preplanned LIPSS or laser-induced irregular surface structures, which might help it overcome the limitations of the existing studies. Here, we are attempting to gather data from laser polarization, laser energy density, scanning speed, and the number of pulses in relation to aspect ratio, groove size, width, depth, spacing, and orientation in different a gaseous environment. The approach has not yet been investigated directly by researchers but some relevant advancements are available, in a variety of materials for a variety of applications.

For instance, the laser assisted nanomachining were already experimented on different materials like metallic elements like K890 cold work tool steel [2], plasma nitrided austenitic stainless steel 304 [3], Cu, Zn, Al, Pb, Mo, Ni, and Fe [6] and transparent dielectrics, single crystalline BaF2 [7], other insulating material, CaF2 (111) [8], transparent glasses, synthetic silica glass ED-H [9], ceramics, titanium nitride (TiN) and diamondlike-carbon (DLC) deposited austenitic stainless steel 304 [10], and semiconductors, InP, GaP, GaAs and InAs [11]. From the different researcher's points of view, the mechanism of nanogrooves fabrication is also different, the Instability of surfaces [7], alteration of refractive index on different medium [12], harmonic generation [11], surface plasmon polaritons excitation [13] and plasma generation [14]. Vibrant and advanced fields of application made us work more on this field, for example, antibacterial effect on 45S5 bioactive glass [15] and 316 L stainless steel [16] controlled cell migration [17] and laser-induced colorization [18] biomimetic surface structuring [19] and superhydrophobicity on AISI 304 stainless steel [20].

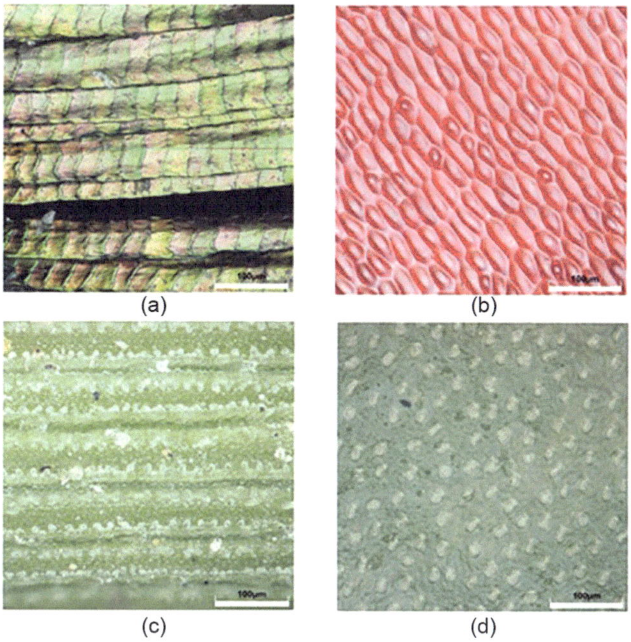

Figure 1.
Different bio surfaces (a) peacock feather (Pavo cristatus), (b) hibiscus, (c) grass leaf (Poaceae), (d) lotus leaf (Nelumbo nucifera).

Nowadays, hierarchical micro - nanostructured surfaces receive considerable attention from researchers worldwide because of their unusual properties [21–24]. Natural or Artificial surfaces with micro-nano surface morphology are crucial in determining the wettability of a solid surface (**Figure 1**).

2. LASER assisted nanomachining

Nanomachining is one the most advanced machining technology to remove nano levels of materials from the workpiece surface by using a sophisticated tool (**Figure 2**). To perform this kind of machining needs an advanced sophisticated tool and nano level resolutioned motion controller.

All of our experiments employed on AISI stainless steel 304, which has a 1.01 mm thickness and a 0.07 mm level of roughness. We adhered to the 10-minute acetone-assisted ultrasonic cleaning procedure, both before and after the trial.

All of our research was conducted using a cutting-edge Ti Sapphire laser system. As shown in **Figure 3**, the system had five axes of motion and 100 fs pulsed lasers with an 800 nm wavelength and a 10 kHz repetition rate. For nanomachining study, we chose a 50 mm focal length and a 25-micron meter spot diameter.

We also took a class of metallic materials to conduct other experiments. There also used ultra-short pulsed laser micromachining system with an advanced workstation. Cleaned with ultrasonicator and etched with suitable etchants. Topography studies were conducted with Opto-Digital Microscope, Stereo Microscope, Scanning Electron Microscope, Field Emission Scanning Electron Microscope, Atomic Force Microscope. Moreover, another methodological characterization was done with Digital Goniometer, UV Visible IR Spectrometer, Microhardness Tester, and X-ray

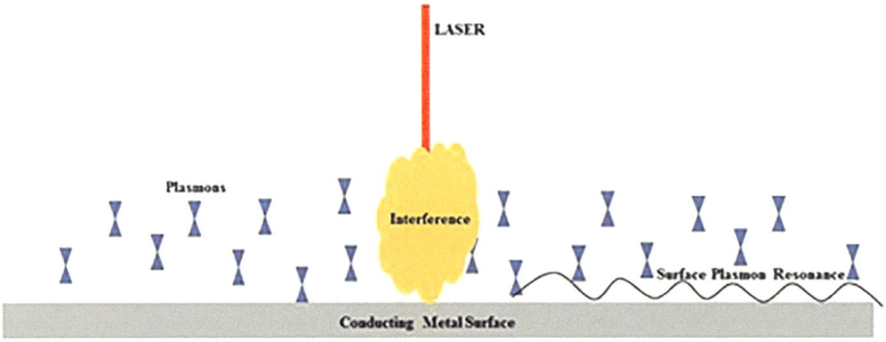

Figure 2.
Interference of LASER with surface plasmons.

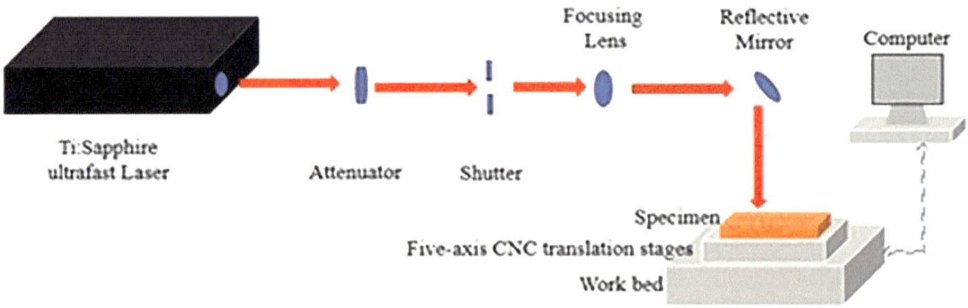

Figure 3.
Laser system with work station.

Photoelectron Spectroscopy. We have selected the Olympus DSX 510 optodigital microscope with a different objective lens, the Hitachi S3400N scanning electron microscope (SEM), and the Park Systems atomic force microscope for the topographical study.

2.1 Nanomachining tuning

2.1.1 Polarization

2.1.1.1 Linear polarization ('↔' Electric field vector)

The effect of linear polarization in nanogrooves generation was studied with a set of experiments in the horizontal, vertical, and inclined direction, represented in the below schematic diagram **Figure 4**. In linear polarization, the grooves are oriented perpendicular in X–direction machining and parallel in Y – direction machining. **Figure 5** represents its processed SEM image and AFM image with line profile, here we evidenced the generation of grooves is in the same format as explained earlier. The fusion effects of electromagnetic interference and its melt hydrodynamics may the conditions of generation of grooves, and the hydrothermal waves from Marangoni shear also influenced [25]. Generally, we cannot easily explain the formation of pattern with the convection or van der walls force performed hydrodynamic instabilities. However, may the temperature instability with a surface profile explain the formation of nano features by evaporation velocity.

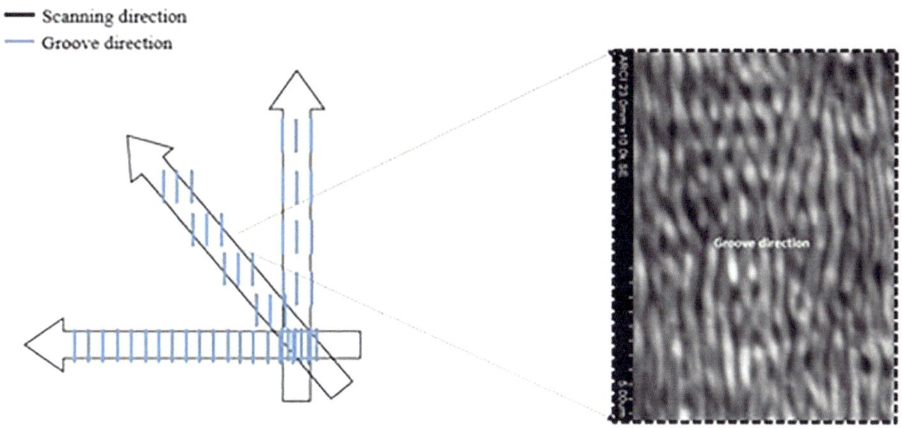

Figure 4.
Schematic representation of machining on its scanning direction with generated grooves (orientation), SEM image of corresponding biomimetic metallic surface [10.0 k SE magnification].

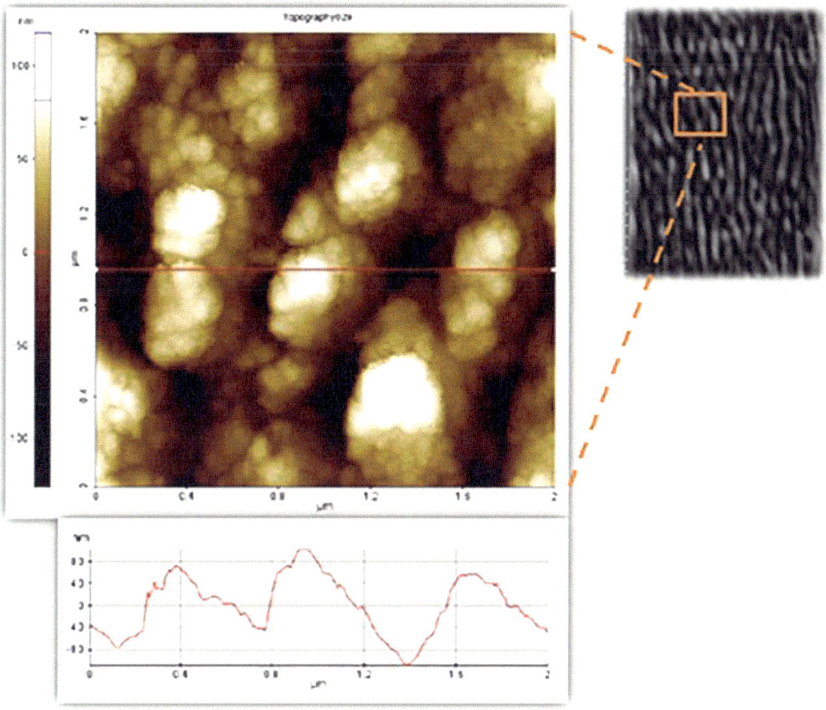

Figure 5.
Processed SEM image [10.0 k SE magnification] and AFM image of 300 nm groove width 100 nm depth [scales are in μm] of surface nanostructures and its line profile for corresponding biomimetic metallic surface.

a. Aspect ratio

Aspect ratio means the ratio of the width of groove and length of groove has studied with the axial distance from the middle of laser path or the center of the laser

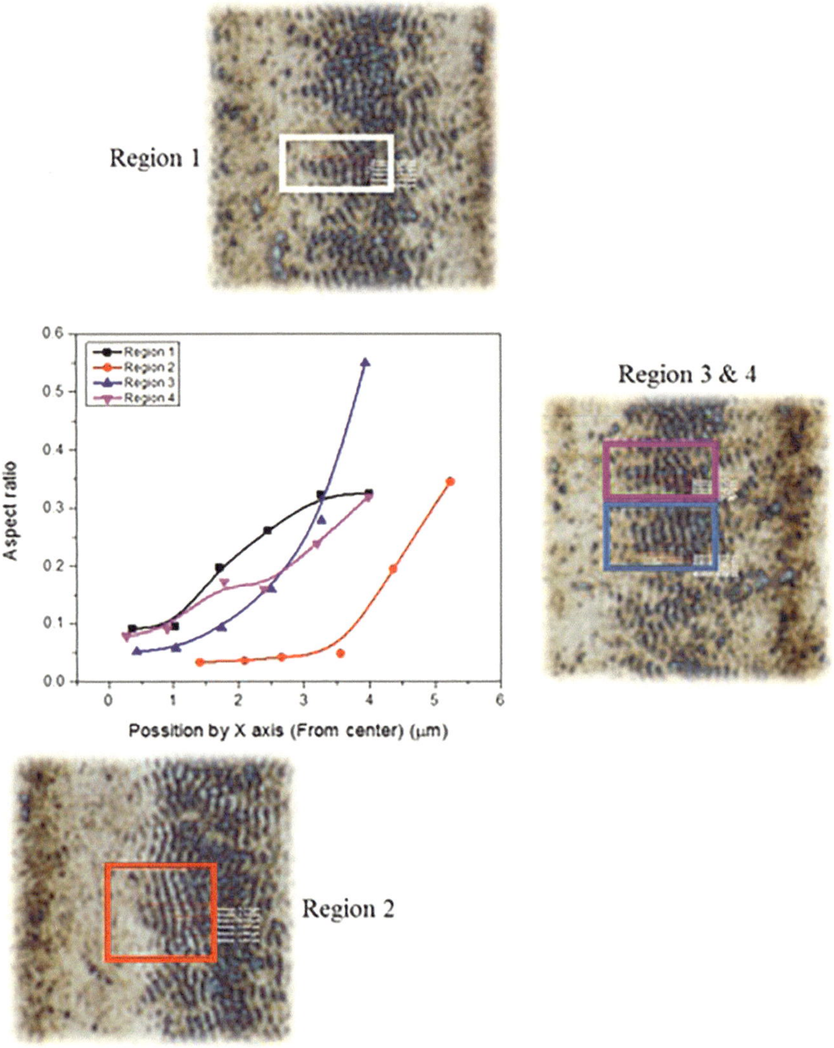

Figure 6.
Aspect ratio of periodic groove structure with axial spacing from the center of LASER spot on biomimetic stainless steel surface [12,481 X OM magnification].

pulse. Result reviles, under high scanning speed (~100 mm/s) of lower energy laser pulses (~40.74 mJ/cm^2) at 10000 Hz, generate a very rich nanostructure with periodically good orientation.

In conclusion, we opted four regions and found an aspect ratio. These four are shown in the above **Figure 6**. Here, the aspect ratio is varied with axial outreach from the center of the laser pulse. The rise of aspect ratio may be by the energy influence, which means that it is getting varied as per spatial Gaussian beam delivery schematically shown in **Figure 7**. May the number of photons for absorption will get altered slightly on aspect ratio and axial distance but the trend remains same [26].

b. LIPSS to laser-induced irregular surface structures transformation

The above-shown **Figure 8** represents the alteration of LIPSS to laser-induced irregular surface structures by the increment of pulses in a specified area. Initially, in two numbers of pulse, the feature is perfectly linearly ordered, and in 10 numbers of pulses also showing somewhat same feature with slight modification under linear polarization. The entire feature transformed to laser-induced irregular surface structures or pillar format in many pulses.

c. laser-induced irregular surface structures to LIPSS transformation

Figure 9 shows the random surface feature to periodic surface feature transformation by rising material interaction energy. Here the input energy is getting riced by the number of multiple thresholds of stainless steel under linear polarization.

In linear polarization (↕ electric field vector), the polarization electric field vector is along the Y-axis and perpendicular to the X-axis, then the laser-induced structures perpendicular to the electric field. The nanogrooves alteration in a particular direction got influenced with the dense electron plasma by multyphoton excitation at the surface. The dense electron plasma generated by the front end of the pulse enhances the remaining electric field and breaks into different nano levels, which may be perpendicular to the direction or parallel to the direction shown in figs.

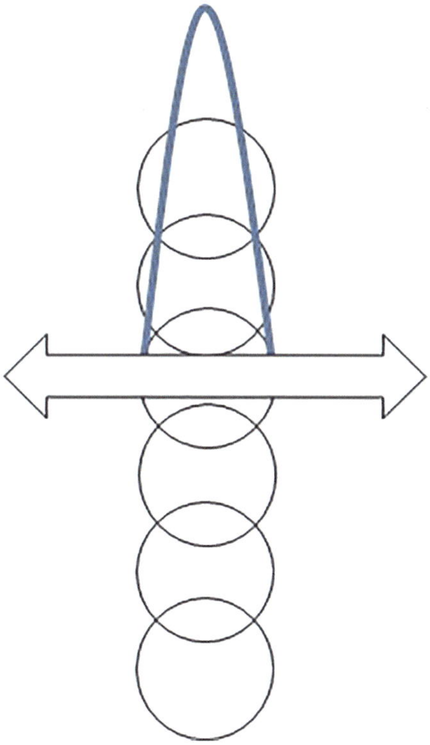

Figure 7.
Schematic representation of laser pulses with energy variance.

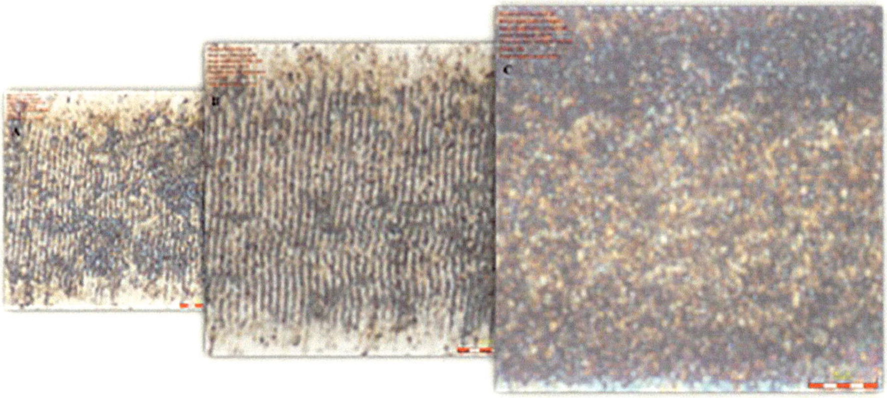

Figure 8.
LIPSS to laser-induced irregular surface structures transformation of biomimetic metallic surface at 40.74 mJ/cm² energy dencity [9707 X OM magnification] (A) 2 pulses, (B) 10 pulses, (C) 100 pulses.

Figure 9.
Laser-induced irregular surface structures to LIPSS transformation of biomimetic metallic surface at 10 number of pulses [9707 X OM magnification] (A) 142.6 mJ/cm² (B) 101.86 mJ/cm² (C) 40.74 mJ/cm².

3. Circular polarization ('O' electric field vector)

For the study of circular polarization, we have exsiccated a single laser path experiment with different angular displacements like horizontal, 45° inclined, and vertical. The analyzed optodigital images incise the variation of nanogrooves.

A circularly polarized laser beam with rounded electric field vector making structure as to not in any particular orientation (**Figure 10**). It depends on the orientation of the impinging moment, may whether perpendicular or parallel or other.

3.1 Nano to micro machining

In circular polarization, the machining from nano to micro depends on the laser energy used. The above-represented **Figure 11** shows the cubic micrometer level volumetric material variation data on different energy densities. Here, it shows the transition of the athermal region to thermal region of material removal. In the above, all LIPSS to laser-induced irregular surface structures study belongs to both optical penetration region and thermal penetration region. Here we explain machining precision from nano level to micro to millimeter level.

In 203.71 mJ/cm^2 energy laser pulses with circular polarization at 10000 Hz frequency generates laser-induced irregular surface structures but different roughness factors. The main influential factor for laser-induced irregular surface structures is high energy or under higher multiple threshold fluence. May the collective upshot of the pressure wave and shock wave causing roughened surfaces and corrosive or non-corrosive working index also alter the feature slightly. The working environment of helium is much clear on microscope images due to its inertness. We have noted no differentiable deviation in orientation due to the accordance of refractive index rice of working environment. The observable difference is that the features' size varies with the working ambiance or the refractive index alteration on different mediums. The periodicity of the nanogrooves reduced. The threshold energy required to generate the LIPSS features on the different working ambiances are different [27], and the laser-material interaction angles are also a matter in these regards." On the delivery of femtosecond laser may the medium drastically influence, or it may alter the focused beam profile in the air than inert gas like helium and argon [28].

In self-cleaning, the surface should be superhydrophobic, and the water droplet should carry a significant amount of dust in isotropic rolling or anisotropic rolling (**Figure 12**).

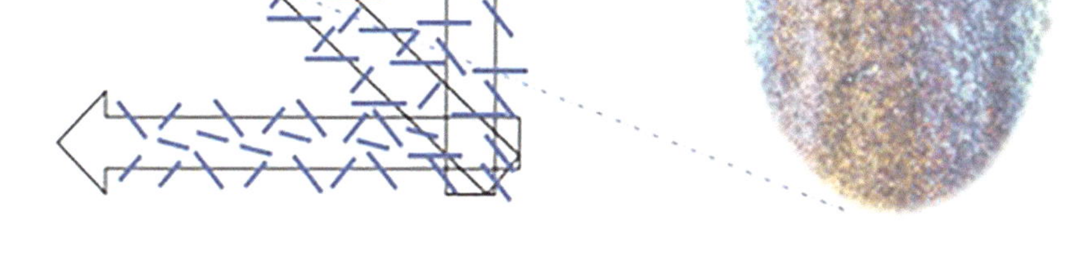

Figure 10.
Schematic representation of machining on its scanning direction with generated grooves (orientation), enlarged focused optical image of corresponding biomimetic metallic surface [~9.0 k OM magnification].

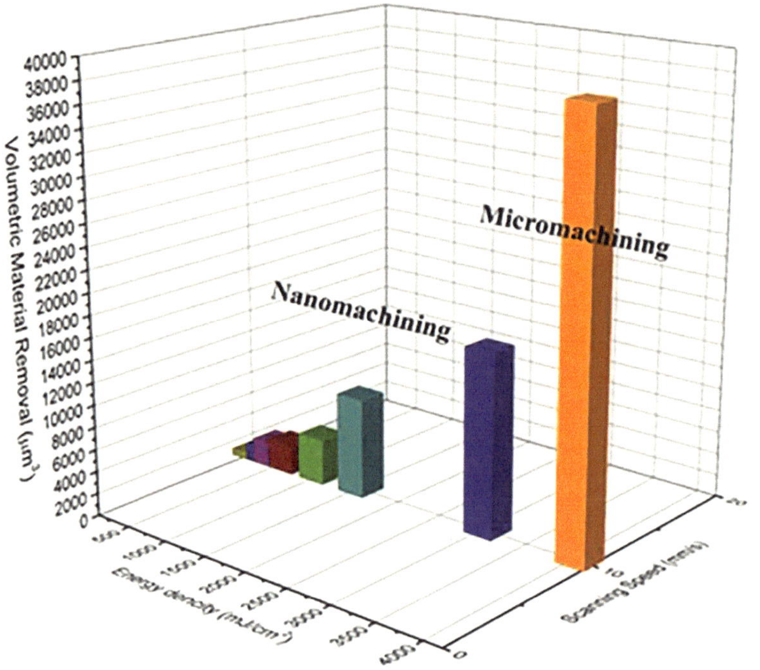

Figure 11.
Representations of material removal rate in (μm³) of different energy dencity at 10 mm/s scanning speed.

The field with directional hydrophobicity is one of the emerging areas to furnish in its full depth. As we know, the lotus leaves with multi-directional or isotropic superhydrophobicity, and grass leaves with unidirectional or anisotropic superhydrophobicity are there to mimic physically [29].

Finding methods to fabricate these complex structures is a vital challenge in surface engineering (**Figure A1**). As we understood the ultrafast laser-assisted micromachining is a method to develop these structures with full functionality. One of our work demonstrates the method and develops multifunctional surfaces with anisotropic and isotropic hydrophobicity using femtosecond laser machining alone. Researchers are already came with anisotropicity of wetting by liquid metals in the past, and the study explained through anisotropicity of interfacial energies on surfaces. The solid, liquid interfaces are much complex to reveal in anisotropic wetting. They demonstrate that the discrepancies occurred while deriving suitabilities in this regard. Hierarchical features by micro and nanostructures are already reliable in hydrophobicity, but the mode of development is getting versatile from time to time. The 130 femtoseconds pulsed laser-assisted hierarchical structure development was also reported earlier, but the mode of machining operations was to be defined. Our study appropriates to place directional hydrophobicity in new advanced dimensions from bio mimics.

4. Conclusion

By the existing knowledge, the nanogrooves generated with interference between the surface plasmon polariton and an incident laser beam means a coupling is

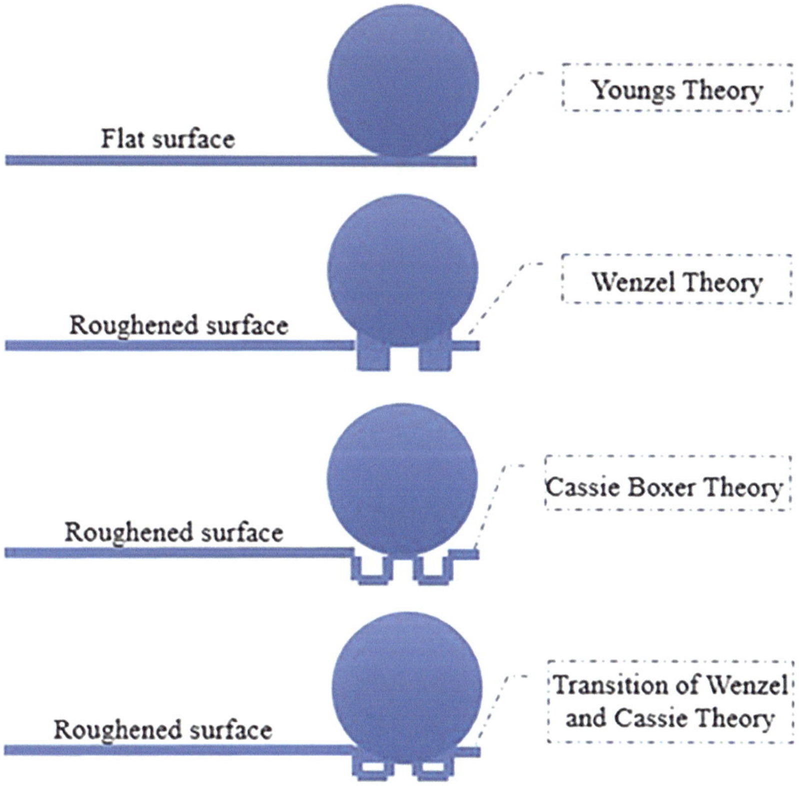

Figure 12.
Theories related to hydrophobicity.

happening with the surface. The surface plasmon directly correlated with surface roughness, so increasing the pulse number will alter the periodicity of the nano surface feature from LIPSS to laser-induced irregular surface structures. The dense electron plasma and its remaining enhancement drive a temperature instability on the normal surface, which directly prompts the evaporation velocity and nano surface feature generation. The rise of aspect ratio was influenced by beam profile and the above-mentioned hydrodynamic temperature instability. The generated surface is getting altered with the working environment also evidenced in this study. Laser energy, number of pulses, the effect of polarization and its angular movement, and normality of topographical material surface are main tuning factors to forge featured surface nanostructure.

We know from the Young Equation to the transition study of the theory of Cassie, Wenzel, continually updating the influential physical attributes to the hydrophobic surface development. Primary surface features and hierarchical surfaces such as two-level hierarchy and three-level hierarchy we demonstrated for our all studies. 30–40 μm sized air trapped hibiscus flower petal surface, Barbs with barbules of peacock feather sized 20–30 μm, 70–90 μm periodic grooves of grass leaves and lotus leaves surface spikes of 20–40 μm are the biological samples we were collected as the imitable evidence. For the optimization study of LASER processing parameters to spectral data, we use Box Behnken Design of experimental study; from these, we establish mathematical relationships for the effects of processing parameters on surface wettability and topography, and it verified with practical data. The BBD methodology allowed for better optimization of ultrafast laser processing parameters for superhydrophobic surfaces. The method with physical alteration will give as the durability of hydrophobic property on the material surface, and the alteration of this property is based on only by another machining process. The method with a fast process for surface alteration is the primary advantage of this applied processing technology on self-cleaning applications. Suitable material with specified texture may provide a foundation to self-cleaning surfaces for food processing components or equipment, hygiene in a food factory that we cannot compensate. It is a very fast method for the development of multifunctional surfaces using femtosecond laser for biomedical application. Hydrophobicity with super or ultra-featured contact angle very low sliding or rolling off-angle can instrument self-cleaning physical activities in solar panels, hydraulic equipments, agri machines for instant residue removal. In addition, to prevent ice formation on power transmission cables and aircraft wings and cockpit shields and corrosion resistance of structures water pipelines hydraulic components and other machineries in addition to biofouling resistance on ship basements structures and other pipelines and thermal barrier coating, enhanced broadband absorption. We have done a proper parameter control on the nanofeature control from nanogrooves to nano spikes.

A. Appendices

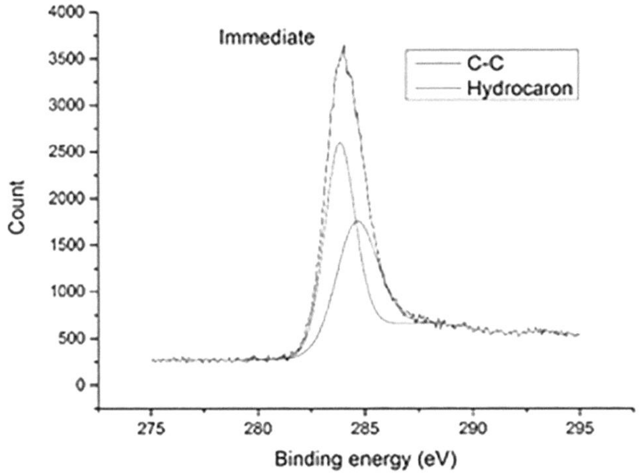

Figure A1.
Elemental analysis (XPS) of modified and unmodified SS 304 surfaces.

Acknowledgements

The researchers, who are noteworthy, produced the manuscripts. Manuscript written by Mr. Srin K S with helpful advice from Prof. Ramkumar J, Dr. Ravi N Bathe, and the Late Dr. G Padmanabham.

All authors acknowledge the assistance of the Department of Science and Technology, Government of India (Grant No: DST/TSG/AMT/2015/628), ARCI Hyderabad, and Indian Institute of Technology Kanpur, India.

Conflict of interest

The authors have no conflict of interest to declare that are relevant to the content of this article.

Thanks

ARCI Hyderabad provided such a relaxable research environment to perform all my experiments to pursue my Ph.D. with all needed facilities. IIT Kanpur provided me the freedom to study all aspects of, which related to my Ph.D. research like a child. Government of India provided me with all the needed helps to pursue my Ph.D. nicely.

Author details

K.S. Srin[1,2]*, J. Ramkumar[1] and Ravi N. Bathe[2]

1 Materials Science Programme, Indian Institute of Technology, Kanpur, India

2 Center for Laser Processing of Materials (CLPM), International Advanced Research Centre for Powder Metallurgy and New Materials, (ARCI), Hyderabad, India

*Address all correspondence to: srin@iitk.ac.in

References

[1] Stephan G, Frank AM. Polarisation-dependent generation of fs-laser induced periodic surface structures. Applied Surface Science. 2015;**331**:150-155. DOI: 10.1016/j.apsusc.2015.01.056

[2] Peter G, Marko S, Bojan P, Jürgen R. Formation of laser-induced periodic surface structures (LIPSS) on tool steel by multiple picosecond laser pulses of different polarizations. Applied Surface Science. 2016;**387**:698-706. DOI: 10.1016/j.apsusc.2016.06.174

[3] Naoki Y, Eisuke S, Kenzo M, Junsuke K. Femtosecond-laser-induced nanostructure formed on nitrided stainless steel. Applied Surface Science. 2012;**264**. DOI: 10.1016/j.apsusc.2012.10.076

[4] Bonsea J, Kruger J, Hohm S, Rosenfeld A. Femtosecond laser-induced periodic surface structures. Journal of Laser Application. 2012;**24**. DOI: 10.2351/1.4712658

[5] Masataka S, Gattass RR, Mazur E. Femtosecond laser-induced formation of nanometer-width grooves on synthetic single-crystal diamond surfaces. Journal of Applied Physics. 2009;**105**. DOI: 10.1063/1.3079512

[6] Hashida M, Fujita M, Tsukamoto M, Semerok AF, Gobert O, et al. Femtosecond laser ablation of metals: Precise measurement and analytical model for crater profiles. Proceedings of SPIE. 2003;**4830**. DOI: 10.1117/12.486538

[7] Reif J, Costache F, Henyk M, Pandelov SV. Ripples revisited: Non-classical morphology at the bottom of femtosecond laser ablation craters in transparent dielectrics. Applied Surface Science. 2002;**197-198**. DOI: 10.1016/S0169-4332(02)00450-6

[8] Costache F, Henyk M, Reif J. Surface patterning on insulators upon femtosecond laser ablation. Applied Surface Science. 2003;**208-209**. DOI: 10.1016/S0169-4332(02)01443-5

[9] Shimotsuma Y, Kazansky PG, Qiu JR, Hirao K. Self-organized Nanogratings in glass irradiated by ultrashort light pulses. Physical Review Letters. 2003;**91**. DOI: 10.1103/PhysRevLett.91.247405

[10] Yasumaru N, Miyazaki K, Kiuchi J. Femtosecond-laser-induced nanostructure formed on hard thin films of TiN and DLC. Applied Physics A. 2003;**76**. DOI: 10.1007/s00339-002-1979-2

[11] Borowiec A, Hauge HK. Subwavelength ripple formation on the surfaces of compound semiconductors irradiated with femtosecond laser pulses. Applied Physics Letters. 2003;**82**. DOI: 10.1063/1.1586457

[12] Wang C, Huo H, Johnson M, Shen M, Mazur E. The thresholds of surface nano–/micro-morphology modifications with femtosecond laser pulse irradiations. Nanotechnology. 2010;**21**. DOI: 10.1088/0957-4484/21/7/075304

[13] Miyaji G, Miyazaki K. Origin of periodicity in nanostructuring on thin film surfaces ablated with femtosecond laser pulses. Optical Express. 2008;**16**. DOI: 10.1364/OE.16.016265

[14] Bhardwaj VR, Simova E, Rajeev PP, Hnatovsky C, Taylor RS, et al. Optically produced arrays of planar nanostructures inside fused silica. Physical Review Letters. 2006;**96**. DOI: 10.1103/PhysRevLett.96.057404

[15] Shaikh S, Deepti S, Sucharita S, et al. Femtosecond laser induced surface

modification for prevention of bacterial adhesion on 45S5 bioactive glass. Journal of Non-Crystalline Solids. 2018;**482**. DOI: 10.1016/j.jnoncrysol.2017.12.019

[16] Lutey AHA, Kling R, et al. Towards Laser-Textured Antibacterial Surfaces. Nature: Scientific Reports; 2018. DOI: 10.1038/s41598-018-28454-2

[17] Martínez-Calderon M, Martín-Palma RJ, et al. Surface micro- and nano-texturing of stainless steel by femtosecond laser for the control of cell migration. Nature: Scientific Reports. 2016;**6**. DOI: 10.1038/srep36296

[18] Maragkaki S, Gurevich EL, et al. Influence of defects on structural colours generated by laser-induced ripples. Nature: Scientific Reports. 2020;**10**. DOI: 10.1038/s41598-019-56638-x

[19] Skoulas E, Stratakis E, et al. Biomimetic surface structuring using cylindrical vector femtosecond laser beams. Nature: Scientific Reports. 2017;**7**. DOI: 10.1038/srep45114

[20] Srin KS, Padmanabham G, Ravi NB. Controllable Superhydrophobic stainless steel surfaces fabrication by femtosecond laser. ASTM: Material Performance and Characterisation. 2019;**8**. DOI: 10.1520/MPC20180090

[21] Wang S, Liu K, Yao X, Jiang L. Bioinspired surfaces with superwettability: New insight on theory, design, and applications. Chemical Reviews. 2015;**115**(16):8230-8293. DOI: 10.1021/cr400083y

[22] Chen F, Zhang D, Yang Q, Yong J, Du G, Si J, et al. Bioinspired wetting surface via laser microfabrication. ACS Applied Materials and Interfaces. 2013;**5**(15):6777-6792. DOI: 10.1021/am401677z

[23] Nine MJ, Tung TT, Alotaibi F, Tran DNH, Losic D. Facile adhesion-tuning of superhydrophobic surfaces between "lotus" and "petal" effect and their influence on icing and deicing properties. ACS Applied Materials and Interfaces. 2017;**9**(9):8393-8402. DOI: 10.1021/acsami.6b16444

[24] Yamamoto M, Nishikawa N, Mayama H, Nonomura Y, Yokojima S, Nakamura S, et al. Theoretical explanation of the lotus effect: Superhydrophobic property changes by removal of nanostructures from the surface of a lotus leaf. Langmuir. 2015;**31**(26):7355-7363. DOI: 10.1021/acs.langmuir.5b00670

[25] George, D, Tsibidis, et al., From ripples to spikes: A hydrodynamical mechanism to interpret femtosecond laser-induced self-assembled structures. Physical Review B. 2015;**92**. DOI: 10.1103/PhysRevB.92.041405

[26] Sugioka K, Cheng Y. Ultrafast lasers—Reliable tools for advanced materials processing. Nature Review-Light: Science & Applications. 2014;**3**. DOI: 10.1038/lsa.2014.30

[27] Gurevich EL, Gurevich SV. Laser induced periodic surface structures induced by surface plasmons coupled via roughness. Applied Surface Science. 2013;**302**. DOI: 10.1016/j.apsusc.2013.10.141

[28] Wang C, et al. The thresholds of surface nano−/micro-morphology modifications with femtosecond laser pulse irradiations. Nanotechnology. IOP Publishing; 2010;**21**. DOI: 10.1088/0957-4484/21/7/075304

[29] Sun J, Longtin JP. Effects of a gas medium on ultrafast laser beam delivery and materials processing. Optical Society of America. 2004;**21**. DOI: 10.1364/JOSAB.21.001081